For President
Griffiths

on the occasion of
his visit
April 3, 1979

THE KREMLIN
AND THE COSMOS

THE KREMLIN AND THE COSMOS

by NICHOLAS DANILOFF

ALFRED A. KNOPF, New York, 1972

THIS IS A BORZOI BOOK
PUBLISHED BY ALFRED A. KNOPF, INC.

Copyright © 1972 by Nicholas Daniloff.
All rights reserved under International and Pan-American
Copyright Conventions. Published in the United States by
Alfred A. Knopf, Inc., New York, and simultaneously
in Canada by Random House of Canada Limited, Toronto.
Distributed by Random House, Inc., New York.
ISBN: 0–394–47493–7
Library of Congress Catalog Card Number: 79–171136
Manufactured in the United States of America.

FIRST EDITION

To the memory of those brave men—
Soviet and American—who gave their
lives in the conquest of space

acknowledgments

This book is a view, by an interested outsider, of Soviet space efforts in the long perspective of history. Undoubtedly it must have its shortcomings, particularly where there continues to be a lack of adequate, open Soviet documentation.

A problem in writing a book such as this is the overabundance of information on some subjects and the paucity of material on others. It may be felt that writing a book on Soviet space efforts from the distance of Washington was foolhardy. But an enormous amount of Soviet material is readily available here in the daily press, specialized journals, transcripts of endless radio broadcasts, popular and scientific volumes. The incomparable resources of the Library of Congress (where much of the research for this book was done) are most usefully supplemented by the files, archives, and libraries of the Department of State, the National Aeronautics and Space Administration, and the Smithsonian Institution and its National Air and Space Museum. In addition, there are a number of individuals in the Washington area who have made a serious hobby of watching Soviet space experiments and who have willingly cooperated with my own efforts.

A brief word about Russian names and titles: In rendering Russian names into the Roman alphabet I have not employed any of the various academic or library systems of transliteration. Rather, I have tried to give them in as simple a form as possible. Where Russian names are quite familiar in the West I have adopted these usual forms. In the bibliography I have translated the Russian titles directly into English rather than give the full Russian versions, which would be difficult for the general reader to understand.

I would like to thank all those who have helped me during the last several years; they include numerous U.S. officials as well as representatives of the Novosti Press Agency of the Soviet Union. Naturally, I have taken full responsibility for all I have written and for the errors I may have unwittingly made. I would particularly like to thank Dr. Charles S. Sheldon II, chief of the Science Policy Research Division of the Library of Congress; Frederick C. Durant III, Assistant Director for Astronautics of the National Air and Space Museum; Dr. Eugene M. Emme, Historian of the National Aeronautics and Space Administration; Kenneth A. Kerst, deputy director of the Office of Research and Analysis for the U.S.S.R. at the State Department; and Konstantin L. Zakharchenko, a former consultant at the Library of Congress, who brought to my attention valuable pieces of open and unclassified information about Soviet activities.

Joseph L. Myler, science reporter, and Stewart M. Hensley, diplomatic correspondent of United Press International, also aided by supplying books, magazines, and helpful advice.

Roderick MacLeish, senior commentator of the Westinghouse Broadcasting Company, was an endless source of encouragement during those dark days which most authors experience when they would rather give up than continue. In a very real sense this volume owes its life to Roderick MacLeish, who saved it more than once.

Mrs. Anna Farrell of Arlington, Virginia, typed the manuscript with admirable patience and perseverance.

Finally, I would like to thank Ashbel Green, managing editor of Alfred A. Knopf, Inc., for his interest and encouragement of this project when it was still a collection of disjointed pages.

Nicholas Daniloff

Washington, D.C.
summer 1971

contents

illustrations

following page 80

Konstantin Tsiolkovsky, Russian theorist of space flight, in 1934.

Friderikh Tsander, Soviet rocket pioneer, in 1913.

The Soviet rocket that launched the first sputniks.

Valentin Glushko, an early Soviet rocket pioneer.

A giant rocket on display in Moscow's 1965 Victory Day parade.

Yuri Gagarin, first spaceman, and Chief Designer Sergei Korolyov.

The augmented Vostok rocket, used to put the latest Soyuz spaceships into orbit.

Premier Nikita Khrushchev speaks to cosmonaut Valery Bykovsky in orbit, 1963.

THE KREMLIN
AND THE COSMOS

prologue:

OCTOBER 4, 1957

In the early hours of Saturday, October 5, 1957, an employee of the Soviet news agency TASS bent over a teletype and flicked a switch. From the TASS headquarters on the edge of tree-lined Tverskoi Boulevard, a message began to clatter out to the world:

For several years research and experimental design work has been underway in the Soviet Union to create artificial satellites of the earth. It has already been reported in the press that the launching of the earth satellites in the U.S.S.R. was planned in accordance with the International Geophysical Year.

As a result of the intensive work by research institutes and designing bureaus, the first satellite was successfully launched in the U.S.S.R. October 4 . . .

In London, because of a three-hour time difference, it was still only Friday night, October 4, and a few minutes before midnight. The international news services, Associated Press, United Press, Reuters, and Agence France Presse, were completing their reports for morning newspapers and broadcasting clients. It had been a particularly quiet night, according to Henry W. Thornberry, the late-night editor for United Press. Early editions of the London morning newspapers contained no major items; there were no impending problems and Thornberry was ready to go home. He had put on his overcoat and sat at the news desk while waiting for the relief editor who would be a few minutes late coming on.

"I could have left the news desk and sat down at the other end of the office," Thornberry recalled years later, "but for some reason I sat at the desk, just doodling aimlessly, on the paper roll of one of the teletype machines. You know which machine it was? The TASS printer. Suddenly, its keys came alive. It didn't take more than a few seconds to realize the significance of what was happening. The Russians had launched the first artificial satellite in the world! Automatically, I began sending off hurried bulletins to New York. I didn't get a chance to take my coat off for two hours, and it wasn't until six o'clock in the morning that I left the office."[1]

In Washington, scientists and newspapermen were gathering in the ornate, gilded ballroom of the Soviet Embassy at 1125 Sixteenth Street, N.W. The occasion was an evening reception offered by the Soviet delegation to a conference on coordinating rocket probes and satellite launches during the International Geophysical Year. The conference was nearly over—it was scheduled to end on Saturday, October 5—and *New York Times* science correspondent Walter Sullivan thought he was onto a major story: in an effort to take continuous geophysical soundings at extremely high altitudes, the United States and the Soviet Union had announced in 1955 that they would orbit the world's first artificial satellites during the year of international scientific cooperation; now, Sullivan believed, the Russians were on the verge of a launching—well ahead of the United States. The story, he thought, was so sensational that he had delayed writing for several days in order to get further confirmation. But at last he had satisfied himself, and submitted it to his office before leaving for the reception at the embassy.

"It was a story that never got published," Sullivan remembers. "I arrived at the Embassy only to be summoned to the telephone by a Soviet official. It was my newspaper calling to tell me of the successful launching of Sputnik and letting me know my story was already out of date. I told several acquaintances and we decided to inform Lloyd Berkner, the coordinator for rockets and satellites under the International Geophysical Year. Berkner clapped his hands to get the attention of the guests and announced: 'I am informed by *The New York Times* that a satellite is in orbit at an elevation of 900 kilometers. I wish to congratulate our Soviet colleagues on their achievement.'

"The Russians," Sullivan told me more than a decade later, "appeared not to have been informed of the event yet. Quite possibly their communications were slower than the news service reports from Moscow."[2]

This, roughly, was the manner in which the news of

Sputnik and the cosmic era burst upon the lives of ordinary people. There were many reactions: jokes, admiration, disparagement, fear. "Maybe we captured the wrong Germans," an American general is alleged to have muttered at one point. "Just a neat scientific trick," outgoing Defense Secretary Charles E. Wilson said. President Eisenhower's administration tried to minimize the event and emphasize that Sputnik brought no new military threat to U.S. security.

But Sputnik *did* bring shock and confusion. Until Friday night, October 4, 1957, conventional wisdom could hold that the Soviet Union was devious, authoritarian, and technologically backward. Now this judgment had been dramatically challenged by a feat which could be observed in the skies and heard on the air waves. What had happened? Why had the United States not achieved this scientific breakthrough first? Was the United States lagging in scientific research? in general education? in missile development? Was U.S. security threatened? Was American diplomacy outwitted? Such questions tortured the minds of many observers, of Congressmen, of officials, of military officers. October 4 started out an ordinary day and ended a pivot of history. Lieutenant General James M. Gavin, who had been associated with American military efforts to develop long-distance rocketry, called the day "a technological Pearl Harbor."[3]

The surprise attack of Sputnik was not to be forgotten easily. The Russians took the occasion to hail the event not merely as an event connected with the International Geophysical Year, but as a first step of man into space: this was the beginning of the road to the cosmos. Soviet officials now date "The Storming of the

Cosmos"—a usual Russian expression—as beginning on October 4, 1957. Other "space spectaculars" came quickly. Within a month Sputnik-2 orbited, carrying into space an experimental dog which was never recovered. Two years later the Soviet Union sent the world's first rocket to the moon; and in 1961, the first human beings into orbit around the earth. Other pioneering "firsts" followed these great achievements. The Soviet Union was the first to launch a woman into space (1963); was first to launch a three-man crew into orbit (1964); was first to allow a space man to leave his ship and walk in the cosmic void (1965); was first to recover lunar samples with automatic devices, and without the direct participation of space men (1970); was first to create an orbiting space station around the earth (1971).

For my part, I was supremely unaware of the cosmic era when I awoke on the morning of October 5, 1957. As an American student in England, I began that day with the pleasant distraction of reading the newspapers. The news from home was tense enough: there was great trouble between President Eisenhower and Governor Faubus of Arkansas over the Little Rock school integration case. A government had fallen in France; there was an epidemic of Asian flu spreading around the world, and the Queen of England was about to begin a tour of the United States and Canada. But the headlines which greeted me had pushed these interesting events into inconsequential corners. *The Daily Express* shouted, SPACE AGE IS HERE; *The Daily Telegraph* announced a bit more moderately, RUSSIA LAUNCHES EARTH SATELLITE; MIDNIGHT REPORT OF ROCKET TAKE-OFF.

It was my first acquaintance with the Soviet space

program, and inevitably I read of more Soviet achievements. I began to ponder these events more seriously when, in March 1958, I wandered more or less by chance into the London bureau of United Press. There I was to begin working as an editor and deskman. Because I knew Russian, I was asked to keep a close eye on the Soviet Union, keeping in mind that one day I might be sent there.

As time passed, I could not avoid wondering about what we, in the West, had come to call the "space race" or "the race to get man on the moon." These are expressions almost never heard in official pronouncements from Russia. Yet one can scarcely deny that a grandiose effort has been in progress in both the United States and the Soviet Union to carry man to the lunar surface and beyond. The effort, I believe, is inevitable because of man's incurable curiosity. What is less inevitable, it seems to me, is that this undertaking must proceed in fierce competition rather than by some negotiated form of cooperation.

Curiously enough, in Russian literature there exists a famous account written before the Russian Revolution of an imaginary expedition to the moon—and it is conceived as an international scientific effort. The author, to whom I shall return, appropriately included an American—named after Benjamin Franklin—in his fanciful crew. The expedition reached the moon in the year 2017 —much as the American Apollo 11 did in July 1969— and dispatched this message back to earth:

We are well and happy and circling the earth in the moon's orbit on the diametrically opposite side. Two of us landed on the moon, traveled across it and made a collection of lunar rock specimens. Owing to the shortage of vital supplies, they had to abandon the moon without making the thorough study desired. Nevertheless, we obtained the following information: the invisible hemisphere of the moon differs in no substantial way from the visible side seen and studied by terrestrial astronomers. Hardly a trace of the atmosphere or water exists. The firmament is semi-spherical, not flattened, black with countless non-twinkling stars. Day and night are 30 times longer than on earth, therefore at night the temperature drops to minus 250°C, while during the day it reaches plus 100°–150°C. No ordinary, stationary plants were found. There is a fairly varied living world. It is a combination of vegetables and animals with chlorophyll in their skins and capable of feeding on inorganic food like most terrestrial plants . . .[4]

Why not? Why should man not cooperate in exploring the universe? Why did it begin in competition? This was a set of questions which has stayed with me for more than a decade as I contemplated the Soviet successes in space while a correspondent in Western Europe, Moscow, and then Washington. The original questions have by now been compounded with others:

• What spurred the Kremlin's interest in space? and what were the forces which started the Soviet space program? What is the Russians' attitude toward the cosmos?

• Why did the Soviet Union take such an astonishing jump on the United States in the early days of space exploration?

• What is the nature of the Soviet space program? Is it simply peaceful exploration? Is it largely aimed at developing terrible new weapons? Or is it a combination of both?

• Who are the men who run the Soviet space program? What is their background? what are their aspirations?

• Was there really a race between the United States

and the Soviet Union to put men on the moon, or was the competition largely the making of the United States following the historic pronouncements of President John F. Kennedy?

• Now that the moon race—if that is what it was—is over, have the chances for Soviet-American cooperation in the costly and dangerous business of exploring space improved?

The Kremlin and the Cosmos is an attempt to answer these questions; it is an effort to probe the Soviet space program in its historical perspective, and, in so doing, to throw some new light on the space race. The book is less a chronicle or a pleasant narrative than an effort to strip away layers of secrecy and uncover some beginnings. Some might argue that it has been a chancy effort because of the relative paucity of available information. I would claim the opposite: the mystery has made the search all the more compelling.

chapter 1

THE CRADLE OF REASON

The roots of Russian interest in space flight are long indeed and stretch back to curious places—for example, to the Gun Club of Baltimore, Maryland. This honorable society was founded during the American Civil War to advance the technology of artillery and further the cause of the Union. When the armies of President Lincoln triumphed, the club's guiding purpose began to fade and its members slipped into a hapless boredom. To combat their lassitude, the club's chief officer proposed and won approval for a dramatic project: to shoot a projectile to the moon by 1870.

The Gun Club of Baltimore, of course, did not really exist—except in the imagination of Jules Verne, the prolific French science-fiction writer of the late nine-teenth century. The moon project was the central theme of his adventure novel *From the Earth to the Moon,* published in 1865 and followed five years later by its sequel, *Around the Moon.* Beyond its popular success, Verne's tale had an extraordinary influence on those serious minds inspired by thoughts of other worlds. Above all Verne was plausible. He respected the established scientific facts and knowledge of an age burgeoning with discovery and invention. And he knew how to combine the elements of science with high adventure.

In the novel *From the Earth to the Moon,* a French adventurer, Michel Ardan, proposed to the Baltimore Gun Club to travel into space inside the projectile to be shot toward the moon. Ardan argued forcefully for converting the spherical projectile into an elongated space ship—and won his argument among the club's doubting members. Then, on a dark December night sometime in the late 1860s, Ardan was fired toward the moon from a gigantic cannon nine hundred feet long which had been imbedded in a hole bored into the hills near Tampa, Florida. Inside the ship two other travelers kept Ardan company: the club's president, Mr. Barbicane, and a Captain Nicholl.

Jules Verne's fictional spacecraft cruised directly toward the moon until it was deflected slightly from its course by a passing asteroid. An observatory in the Rocky Mountains followed its progress and sighted the craft swinging around the far side of the moon and out of view. It reappeared again in due course, and in a scene uncannily reminiscent of the U.S. Apollo moon missions, headed for a soft landing in the Pacific Ocean. There it was rescued by the U.S.S. *Susquehanna.*

Verne's science fiction won a wide audience, was read not only in the United States and Europe but even in the remote countryside of Tsar Alexander III's Russia. Konstantin Eduardovich Tsiolkovsky, a half-deaf, self-taught Russian school teacher, who has come to be acknowledged as the first great theoretician of inter-planetary flight, credited Verne's adventure stories with setting his own scientific thoughts in motion. "It seems to me," he was to write in 1911, "that the first seeds of the idea were sown by that great science-fiction author Jules Verne. He startled my brain. He directed my thoughts along certain lines; then came a desire, and after that, the work of the mind."[1]

Verne's novels were not the sole source of Tsiolkov-sky's inspiration, however. The Russian was the bene-ficiary of a long stream of intellectual and scientific activity dating back to the second-century Roman writer Lucian of Samosata and to Chinese warriors of the tenth century. Lucian wrote, in his *Vera Historia*, what is believed to be the world's first fantasy about a trip to the moon; the Chinese contributed the first known rockets. Modern astronomy, fireworks, and rocketry de-veloped in Europe during the sixteenth and seventeenth centuries; the eighteenth and nineteenth centuries saw important new scientific strides, and various fantastic tales of other worlds were written.* In Russia, as else-where in nineteenth-century Europe, rockets were being

*Among them are Ralph Morris's *A Narrative of the Life and Aston-ishing Adventures of John Daniel* (1751), Miles Wilson's *Man in the Moon* and *The History of Israel Jobson, the Wandering Jew* (?), Joseph Atterlay's *A Voyage to the Moon with some Account of the Manners and Customs, Science and Philosophy of the People of Morosofia and other Lunarians* (1827), and Edgar Allan Poe's *Lunar Discoveries, Extraordinary Aerial Voyage by Baron Hans Pfaall* (1835).

adapted for military uses, and inventors were putting forward schemes for lighter- and heavier-than-air machines. Historians have documented at least ten different proposals for flying devices in Russia powered by primitive reactive-propulsion engines.[2]

Probably the most memorable of these primitive devices was a flying platform powered by a rocket engine that fired powder and air toward the ground. It was conceived by a twenty-seven-year-old revolutionary, Nikolai I. Kibalchich, who was executed April 3, 1881, for having participated in the assassination of Tsar Alexander II—in fact, for having concocted the bomb which exploded under the Tsar's carriage in St. Petersburg. The police confiscated Kibalchich's calculations, which they considered potentially dangerous, and buried them in their archives to be rediscovered only after the Russian Revolution. The young revolutionary's plans for the flying platform were published in the magazine *Bygone Times* (*Byloe*) in 1918 to the delight of the growing fraternity of space-flight enthusiasts.

Kibalchich's idea had its effect on later generations although not on Tsiolkovsky. Tsiolkovsky had been influenced earlier by a project proposed by A. P. Fyodorov in his pamphlet *A New Principle of Flying*, published in St. Petersburg in 1896. The teacher considered Fyodorov's idea for a jet flying device inexact because it lacked scientific calculations; but it gave him a shock. He compared it to the apple which fell on Sir Isaac Newton's head.[3] Tsiolkovsky's education had undergone other influences. There was the wonder of seeing escape into the sky a hydrogen balloon which his mother gave him when he was eight. There were the after-effects of serious illness a year later. While sledding, Tsiolkovsky contracted a heavy cold, which turned into scarlet fever and

eventually left him almost totally deaf. In a later moment
of reflection, Tsiolkovsky recalled: "In childhood deafness chained me to indescribable tortures, as I was very
curious. Then slowly I became accustomed but it never
really stopped bothering me (even though I recognize
that the originality of my work was precisely due to
it)."[4]

A strong curiosity added to these initial stimuli developed into an obsession with freeing man from the
force of gravity. How wonderful, Tsiolkovsky thought
as a teenager, if man could achieve a weightless state.
He pursued this quest vigorously when his father, a for-
ester by profession, sent him to Moscow to be educated.
Despite his deafness, Tsiolkovsky spent three years
there alone, studying by himself. Existing primarily on
bread, water, and a few staples, he delved unsystemati-
cally into mathematics, physics, and chemistry. He was
sixteen years old, rumpled, half-starved, and considered
by many to be crazy. At one point, Tsiolkovsky believed
he had conquered his obsession with weightlessness by
inventing a device that could defy gravity. He was mistaken, but his conviction left a profound impression:

"I was so agitated, nay shaken, that I could not sleep
that night—just wandered in the streets of Moscow and
thought about the grave consequences of my invention.
But by morning, I understood its futility, and the disillu-
sionment was as great as the illusion had been. That
night left a lasting impact on my life. Now after 30
years, I sometimes dream about travelling to the stars in
my apparatus, and my heart swells with the exultation I
experienced on that unforgettable night."[5]

Tsiolkovsky returned from Moscow in 1878 and settled in Borovsk, Kaluga Province, where he obtained a
high school teaching license. Though he spent much

description explaining how a rocket could be fired toward the moon, setting off a charge of flash powder on impact. His suggestion was met with incredulity and disdain. Goddard was a secretive, reclusive man. The unwanted publicity and its critical reaction apparently caused him to repress his thoughts about interplanetary flight and to engage in serious but ever more solitary rocket studies. Goddard flew the world's first liquid-fueled rocket in Worcester on March 16, 1926, then later moved to Roswell, New Mexico, to continue his experiments with help of foundation grants. Interestingly enough, Tsiolkovsky's 1903 work, which was completely independent from the Goddard study, also focused on the problem of reaching the highest levels of the atmosphere. Tsiolkovsky pointed out that a balloon was capable of rising only to a limited altitude, and that investigations of the upper layers of the atmosphere would require an altogether different kind of propulsion. The Russian theoretician went on to demolish the idea of Jules Verne's giant cannon in the Florida hills, explaining that the shock of sudden acceleration would kill human passengers and that, in any case, the projectile would never attain "escape velocity." Furthermore, Tsiolkovsky asserted, it was impractical to cast a cannon as long as the Eiffel Tower was tall. Instead, the Russian proposed a liquid-fueled rocket ship which would burn liquid oxygen and liquid hydrogen. In later years, he considered more exactly the obstacles to interplanetary flight posed by the earth's gravity and the resistance of the atmosphere. He concluded they could be overcome. He conceived of the notion of rocket stages, or "rocket trains" as he called them, to aid acceleration. He discussed the relative merits of known fuels and suggested the modern method of cooling the rocket's combustion chamber with its own cold, liquid fuel. Tsiolkovsky also perceived the danger of spaceships burning up during re-entry from outer space unless especially protected by a heat shield. What Tsiolkovsky did not do in 1903, or later, was to couple his theories with practical experimentation. Unlike Goddard in the United States, Tsiolkovsky did not build rockets, test them, or improve them.

"In a number of instances," he said, "I have been forced to guess and suggest. I am not deceiving myself and know well that I have not solved the problem in full, but that it is still necessary to work on it 100 times more than I have. My aim is to awaken interest, having shown its [space flight's] great signficance in the future and the possibility of its solution."[7]

Tsiolkovsky did awaken interest, but by no means right away. For one thing his study was so technical that when it appeared in *Scientific Review* few could grasp its meaning. Also the magazine had only a limited circulation. It was only in 1911, when an enlarged version of the article was published in the important *Herald of Aeronautics,* that Tsiolkovsky began to attract attention.

He came to greater prominence after the Bolsheviks seized power in 1917 and Russia withdrew from the First World War. Lenin made vigorous efforts to promote science, and Tsiolkovsky, the neglected, prerevollutionary scientist, became a convenient symbol of Soviet progress. He was elected a member of the Socialist Academy of Sciences in 1919; in 1921, Lenin and the Council of People's Commissars granted him a lifelong pension for his contributions to aviation;[8] after the Revolution his books were freely published.

Rocket flight was now being explored by enthusiastic amateurs in many countries, particularly in Germany.

Professor Herman Oberth published his landmark work, *The Rocket Into Planetary Space* in 1923. In 1926, Goddard began testing liquid-fueled rockets. In 1930, other American enthusiasts formed the American Interplanetary Society and carried out their own tests independently of Goddard. In Russia, Tsiolkovsky had emerged from the status of crank to socialist respectability, and conceived a fourteen-point program for the conquest of space.[9]* "The earth," Tsiolkovsky told his followers, "is the cradle of reason, but one cannot live in a cradle forever."[10] The followers hardly needed convincing; particularly not Friderikh Arturovich Tsander.

A Latvian engineer, Tsander was infected with the same obsession as Tsiolkovsky: interplanetary flight. Furthermore, Tsander did what Tsiolkovsky did not; he experimented with spacecraft construction and tried in a practical way to harness rocket power. And yet, Tsander's greatest contribution was in neither of these two fields of research and experimentation, but rather in the popularization of space flight and in organizing its early enthusiasts into a concentrated, working group. Despite ill health and a lack of official support and money, Tsander clung to his obsession with space flight and infected others. He proclaimed his determination with the exotic slogan "Forward to Mars."

---

*Tsiolkovsky's plan included these stages: 1. development of a rocket plane with wings; 2. aircraft further developed with shorter wings; 3. developed to attain altitude of 12 kilometers; 4. wingless vehicles developed; 5. rocket developed that is capable of speed of eight kilometers per second; 6. first flights into the cosmos; 7. development of regenerative processes in the cabin; 8. spacesuits developed; 9. plants carried into space to aid regenerative systems; 10. space stations set up around the earth; 11. solar energy harnessed for space locomotion; 12. colonies established on asteroid; 13. colonies developed further; 14. "Human society and its individual members become perfect."

Like Tsiolkovsky's, Tsander's interest in exploring the solar system was the product of several urges. Of these the most obvious was Tsiolkovsky's 1903 treatise, *Exploring Cosmic Space with Reactive Devices*. Tsander became acquainted with it as a high school student when his astronomy teacher read it in class before the Christmas vacation of 1904. Jules Verne's *From the Earth to the Moon* and *Around the Moon* also became lifelong favorites.[11]

Tsander was born in Riga on August 24, 1887, and lost his mother when he was two years old. His father, a doctor employed at the Riga Zoological Museum, participated actively in his son's education. Frequently, the two visited the Riga museum. Tsander said later that these visits had a strong effect on him. Seeing strange animals, meteorites, and astronomical models "sparked in me from early childhood the desire to fly to the stars," he wrote. He was fascinated by the experiments of the German inventor Otto Lilienthal who, toward the end of the nineteenth century, studied bird flight and constructed gliders demonstrating the advantages of curved surfaces over flat ones for wings.[12]

By the time he graduated from the Riga Polytechnical Institute on July 31, 1914, Friderikh Tsander was totally engrossed in the problems of space flight. Recent research has uncovered his "Spaceship Notebook," which he started in 1907, when he was twenty, in order to develop ideas for the construction of spacecraft. The following year he organized the Riga Student Society of Space Navigation and Technology of Flight. One of his practical experiments in the early years of the century was to build a lightweight greenhouse for a spaceship. It was fertilized by what he called "night gold," or human excrement.[13] Tsander had, in other words, worked out a primitive regenerative system, whose essential technique is used in today's space flights.

When, between the two World Wars, Latvia became independent, Tsander left Riga. He is said to have favored the Communist Party although he never became a member. In any case, he arrived in Moscow to enter the Soviet aviation factory Motor No. 4 in 1919. There he became chief of the technical construction and statistical bureau. His professional duties involved the development of aircraft engines while his personal preoccupation remained the use of aircraft for reaching high altitudes. Much of Tsander's space interest during this period is recorded in personal files that his family recently donated to the Soviet Academy of Sciences. The files amount to over five thousand pages and are difficult to decipher because Tsander made notes in an archaic form of German shorthand.[14]

He, like Tsiolkovsky, ultimately received the attention of the highest Soviet authorities, although his encounters never resulted in financial help. Tsander did have a brief meeting with Lenin during which they discussed the possibility of interplanetary flight. Tsander has described the meeting, which occurred after a lecture he gave for a group of Moscow inventors in early January, 1920:

"Before the lecture I was informed that Lenin would be in the audience. This upset me and I became nervous.

"After the speech, I was invited to meet Lenin; this made me confused. Lenin was greatly interested in my work and my plans for the future; he spoke with such simplicity and cordiality that I am afraid I took advantage of his time by relating to him in great detail my work and my determination to build a rocket spaceship.

"I also told Lenin that I was working on the problem of man's flight to Mars: the construction of a suitable spaceship, which methods to devise in order to assist man to overcome the acceleration, and also the question of suitable clothing and diet.

"Lenin asked me, 'Will you be the first to fly?' I answered that I had to set an example and that I never thought possible to do otherwise. I was sure others would subsequently fly after me. At the end of our conversation, Lenin shook my hand strongly, wished me success in my work, and promised support.

"Lenin made a tremendous impression on me. That night I could not sleep. Pacing up and down in my small room, I thought of the greatness of this man—our country is ravaged by war; there is a lack of bread, of coal, and the factories are at a standstill, but this man who controls this huge country finds time to listen to space flights. It means my wishes will come true, I thought."[15]

Tsander's enthusiasm is oddly reminiscent of Tsiolkovsky's joy when, as a sixteen-year-old, the patriarch of Russian space flight thought he had invented a device which could conquer gravity. In 1922, Tsander took a year's leave from his factory job to work on designing a spaceship. He also tirelessly popularized the notion of space flight at a time when Russia was wracked with economic problems following the bitter civil war.

Tsander participated in a debate on October 1, 1924, which was called to discuss a report that Professor Goddard had attempted to send a rocket with flash powder to the moon during that year's Fourth of July celebrations at Worcester, Massachusetts. The report, in fact, was erroneous, but it had gained considerable circulation in the Western press and was noticed in Russia. The evening discussion caused so much excitement that it had to be continued on October 4 and again on October 5 with mounted police to keep order.

Three years later, in connection with the tenth anniversary of the Russian Revolution, the Moscow Association of Inventors held a space exhibition that Soviet histories now hail as the first of its kind in the world. The exhibit ran from April to June, 1927, and displayed spaceships designed by Tsiolkovsky, Tsander, and others. There was even a corner devoted to Goddard and Oberth, and the organizers took care to honor Jules Verne and the English writer H. G. Wells. Reading over some of the records of the exhibit, one is impressed with the fascination it stirred up. Salomei G. Vortkin, a writer for the newspaper *Rabochaya Moskva,* wrote in the visitor's book: "I want to fly with you on the first flight. My desire is serious. As soon as I hear that you are ready, I shall try with all my might to see that you take me too. I pray you will not put obstacles in the way of achieving my desire."[16]

The Moscow organizers of the exhibition may have been inspired by this kind of support. But reality was starker. The Moscow press was skeptical; the authorities declined to give the organizers adequate financial support. In 1927, economic hardships continued. Lenin had been dead for three years and Stalin was initiating a critical struggle for power with Leon Trotsky.

It is not really surprising that the working-level authorities were dubious about space flight in the 1920s even though Lenin recognized its theoretical importance. What Tsander and the other amateurs were proposing was still well beyond the reach of contemporary technology. Tsander, whose enthusiasm had moved him to name his son Mercury and his daughter Astra, was proposing a spaceship which would consume its stubby wings in outer space by converting their metal into highly calorific fuel. He displayed a model of his craft at the 1927 exhibition, and went on to suggest that eventually solar energy could also be harnessed to propel space vehicles.

Tsander unsuccessfully pressed Soviet authorities at various levels to permit him to work full time at state expense on problems of space flight. He sought help from the military by dispatching a letter to Defense Commissar Kliment Ye. Voroshilov; he approached Education Commissar Anatoly V. Lunacharsky. He pleaded that if he were given the facilities of Aviatrest and the Zhukovsky Central Aero-Hydrodynamic Institute (TsAGI)—two institutions concerned with aircraft development—his efforts would amount to a mere 1/5000th of Soviet investment in aviation.

But Glavnauka, the Soviet agency coordinating scientific research, dispatched its refusal to him July 7, 1927. "It is not considered possible to satisfy your petition for assistance in completing your work on interplanetary travel."[17] Undaunted, Tsander began designing his own rocket engines, the first of which was the OR-1, fashioned from an old blowtorch.

The 1920s and '30s mark an important transition in rocket development in Russia, Europe, and the United States. They were a period when rocket propulsion became less a theory and more a practical reality. Following Dr. Goddard's rocket experiments toward the end of the 1920s, rocket clubs and societies proliferated in the United States. In Germany, Professor Oberth became president of the Verein fur Raumschiffarht (Society for Space Travel), which had been founded in 1927. The society members experimented with rockets to power gliders and autos. Eventually some of its members would make significant contributions to the German army-rocket program that gave birth to the V2, which ballistic missile was used by Hitler to bombard Great Britain in 1944.

In May 1931, a group of Moscow scientists pushed for the formation of a special group to study rocket propulsion within the framework of OSOAVIAKHIM, the paramilitary society for the promotion of defense, aviation, and chemical industries. A jet propulsion section of the society was founded and eventually became known as the Group for the Study of Reactive Propulsion: GIRD.[18] Tsander was a leading figure. The group developed branches in various cities throughout the Soviet Union, particularly in Leningrad. Its members worked on the creation of rockets and rocket engines of various descriptions.

In a renovated Moscow basement, Tsander's section toiled in their spare hours, experimenting and designing rocket engines. Tsander ran some fifty combustion tests of his OR-1 engine between 1929 and 1930, proving that it worked. Having done this, he started on the OR-2, a liquid-fuel engine that ultimately developed enough thrust to power theoretically an experimental aircraft. By 1933, the Moscow group was working on four liquid-fuel rockets, one of which, the 09, was the first Soviet liquid-fuel rocket to fly. Sergei P. Korolyov, a young engineer who headed the testing team, recorded the historic event on August 17, 1933:

We, the undersigned, a committee of the GIRD factory for the launching of an experimental prototype device 09, including:

—GIRD leader, civilian engineer, Korolyov, S. P.,
—Civilian engineer of brigade No. 2, Yefremova, N. I.,
—chief of brigade No. 1, civilian engineer, Korneyev, L. K.,
—brigade member, welder, Matysaka, E. M.,

on this seventeenth of August, having examined the device and the launching apparatus, ordered it launched into the sky.

The launch occurred at station No. 17 at the engineering sector of Nakhabino, 17 August at 1900 hours.

Weight of the device—18 kilograms.
Weight of fuel-gasoline jelly—1 kilogram.
Weight of oxygen—3.45 kilograms.
Pressure in the oxygen tank—13.5 kilograms.
Length of flight from the moment of launch until moment of fall—18 seconds.
Height of vertical rise (by eyesight)—about 400 meters.

Lift-off took place slowly. At maximum height, the rocket turned over into a horizontal plane and then fell through an arched trajectory into a neighboring wood. During the fall to earth, a mantle crumpled.

The change from vertical flight to horizontal and then the dive to earth took place as the result of gases burning through one flange, in consequence of which a lateral force arose which also overturned the rocket.

Done in one copy and signed at the Nakhabino launching pad, 17 August, at 2010 hours, 1933.[19]

On November 25, 1933, another rocket—designated the GIRD-X—followed successfully into the air. But

Friderikh Tsander did not live to see it. Ailing from overwork, he had been persuaded to leave Moscow for a rest cure at the mineral spa at Kislovodsk. On the long train trip he contracted typhoid fever, and died sometime after his arrival on March 28, 1933. Korolyov, who succeeded him at the head of GIRD, announced the news at a somber meeting recalled by one member: "We all gathered in one of the large GIRD rooms. S. P. Korolyov came up to the podium and reported to us the death of Friderikh Arturovich. A deep silence reigned. Here for the first time I saw tears on his (Korolyov's) severe, intelligent, strong-willed face. Our sad silence lasted for a long while."[20]

Tsander was gone; but an important legacy remained —the organized enthusiasts of Moscow and of Leningrad. It was in the spring of 1931 that the planning of the Leningrad GIRD got under way. Vladimir V. Razumov, subsequently elected chairman, has left a fascinating account of the formation of this group; it reveals the rather modest financial resources of an important jet study society and it shows how this comparative poverty led to the development of close ties between the early rocket experimenters and the Soviet military establishment.[21]

The Leningrad group, known as LenGIRD, held its first formal meeting in November 1931 and elected officers. Razumov became chairman, and Dr. Yakov Perelman, a major popularizer of Tsiolkovsky, vice-chairman. Also elected an officer was Professor Nikolai A. Rynin, the author of an impressive nine-volume space flight encyclopedia published in 1929. In its early days, LenGIRD consisted of only about forty members, who devoted all their spare time working in their own cramped quarters, sometimes in the apartment of Razumov himself. The members' limited means stimulated the urgent search for money, which led them to the People's Defense Commissariat.

Razumov relates in his account how he went to Moscow shortly after the formation of LenGIRD to petition Mikhail N. Tukhachevsky, Deputy Defense Commissar, who eventually became Red Army Chief of Staff. Tukhachevsky, executed during the purges of the late 1930s on trumped-up charges of being a Nazi spy, conferred with Razumov and expressed interest in the military possibilities of rocket development. To encourage the Leningrad enthusiasts, he arranged a grant of fourteen thousand rubles for the construction of a high-altitude rocket. This was equivalent to about two thousand dollars, and Tukhachevsky promised more funds for the future.[22]

Besides LenGIRD, another Leningrad institution played a crucial role in the development of Soviet rocketry. This was the Leningrad Gas Dynamics Laboratory (Gazo-Dynamicheskaya Laboratoriya), whose special construction bureau was eventually to design and develop the powerful rocket engines that drove the Sputnik booster in 1957 and perfected the first stage of the rockets that launched all of Russia's cosmonauts into space through 1971.

Known as GDL, this laboratory was formally established as the Military Scientific Research Committee of the Soviet Armed Forces. It traced its origins back to two Moscow engineers, Nikolai I. Tikhomirov and his assistant, Vladimir A. Artemev. Tikhomirov and Artemev were first established in Moscow in 1921 and were concerned with the development of solid-fueled rockets in the tradition of the Russian military going back to the nineteenth century. They subsequently moved to Leningrad to be closer to Soviet artillery authorities, who in the late 1920s were interested in solid-fueled rockets. Boris S. Petropavlovsky, a military engineer and avid space flight enthusiast, succeeded Tikhomirov as the head of the Gas Dynamics Laboratory. Through Petropavlovsky informal links began to develop between GDL and LenGIRD.[23]

These close ties between GDL and LenGIRD were consolidated by the new laboratory director, Ivan T. Kleimenov, who succeeded on the death of Petropavlovsky in 1933. Indeed, Kleimenov suggested to the military that all rocket work by military and civilian specialists in Leningrad and Moscow should be unified into a single institution. Kleimenov's proposal was backed by Tukhachevsky, who on occasion had witnessed demonstrations of military rocket power put on by GDL. Grigory K. Ordzhonikidze, a member of the ruling Politburo of the Communist Party, also blessed the Kleimenov proposal; and, at the end of 1933, Tukhachevsky signed the decision setting up a Jet Scientific Research Institute (Reaktivny Nauchno-Issledovatelsky Institut). The Soviet Union thus acquired, beginning in 1934, a national institution concerned with developing solid-fuel and liquid-fuel rockets for military purposes. It took another eight years for the United States military to give similar backing to rocket development.

The Jet Scientific Research Institute (RNII as it was called in Russian) continued to take an interest in space flight although military applications of rocketry remained a primary concern. Under Kleimenov, the Gas

Dynamics Laboratory had opened a "Second Section" devoted to the practical development of the jet propulsion theories of Tsiolkovsky.[24] Indeed, after the formation of the Jet Scientific Research Institute, Kleimenov sent this letter to the venerable pioneer:

7 February 1934

Dear Konstantin Eduardovich:

At the end of 1933, by a decision of the Government the separate organizations and groups working on the problem of jet propulsion were amalgamated into the Jet Scientific Research Institute, which is responsible for the development and construction of aircraft whose motion is governed by the jet principle.

The dream of all researchers in this new field of human knowledge has thus been fulfilled: We have a base for tremendous development, from their scientific beginnings, of those ideas whose first herald you were.

There is no doubt that the organization of this Institute has been made possible only through the conditions created by the struggle of the many millions of Soviet working classes, under the leadership of the Communist Party, for Communist reconstruction of all human society and for conquest of the heights of science and engineering.

We consider it indispensable to keep in close touch with you, the creator and contributor of the foundations of the theory of jet propulsion, and ask your consent to your appointment, in the near future, as one of the three or four leading workers of our Institute.

May we request a prompt reply by telegram.

With friendly greetings,

I. Kleimenov.[25]

Shortly afterward, when formally notified of his election, Tsiolkovsky replied to Kleimenov:

9 March 1934

Dear Ivan Terentevich:

I have just received your letter and am answering it the same day.

I hope by my work to show my gratitude to you and to all the workers of the Jet Scientific Research Institute.

I am sending you herewith an article on explosives. Forward it to the journal, unless you feel it to merit secrecy, in which case I have no objection, as long as it is useful to the work than which nothing loftier has yet appeared on earth.

Regards to Tikhomirov and to the whole Institute.

K. Tsiolkovsky

The need for secrecy is stressed even in this letter of 1934; extensive secrecy was to engulf the Soviet space program in the 1950s and '60s. The Russian space flight pioneers were not prolific recorders of their achievements in the period before the Second World War. It is true, of course, that Westerners gained some insight into Soviet developments through a few early Soviet rocketeers who became members of foreign jet-propulsion societies.[26] But with the formation of the Jet Scientific Research Institute a screen of military secrecy began to descend. Tsiolkovsky, who occasionally corresponded with scientists abroad, died on September 19, 1935. Shortly before his death he exchanged warm messages with Stalin and willed his research and papers to the Communist Party[27]—where secrecy was a way of life.

Time has revealed some of the achievements made by the Jet Scientific Research Institute. Under the direction of Valentin P. Glushko, a scientist who reappears in the 1950s and '60s as a leading rocket-engine designer, the Gas Dynamics Laboratory developed a celebrated series of liquid-fuel engines called the ORM (Opytny Raketny Motor—Experimental Rocket Engine) group. Korolyov, whose fame also blossomed after Sputnik, designed a winged missile, roughly similar to the V1 pulse jet weapon used by the German Luftwaffe during the Second World War. The Soviet Air Force used primitive air-to-air rockets against invading Japanese aircraft in Mongolia in 1939; Korolyov designed a glider powered by a rocket engine which flew on February 28, 1940; aircraft designer V. F. Bolkhovitinov created a rocket-powered interceptor, the BI-1, which flew on May 15, 1942; the Jet Scientific Research Institute developed rocket units to assist the takeoff of military aircraft during World War II, while Soviet forces used batteries of "Katyusha" rockets with devastating effect against the Nazis at Stalingrad and in other battles.

It was only after the war that the United States and Great Britain grasped the advancing state of Soviet rocketry. In the late 1940s Allied intelligence agents obtained, by accident, a general report on Soviet rocket developments to date—and suddenly the West realized it had a lot of catching up to do.

chapter 2

THE DREAM
AT WAR

Twenty years after the Nazi invasion of the Soviet Union there were few physical reminders in Moscow of the colossal devastation which the Russians suffered during the Second World War. But there were deep psychological scars.

On November 25, 1961, President John F. Kennedy had granted, after months of elaborate arrangements, a lengthy interview with the then editor of the Soviet government newspaper *Izvestia,* Alexei I. Adzhubei. Considerable attention was focused in the United States on the question of whether the Soviet authorities would fulfill their agreement to publish the interview in full, thus presenting to millions of Soviet readers a full account of Kennedy's views on world affairs. On November 29, a cold, snowbound night in Moscow, foreign correspondents picked up the *Izvestia* text for immediate translation and transmission to the West. Their first professional concern was: Is this the complete, unadulterated interview? They concluded that it was complete except for a minor rephrasing of one of Adzhubei's questions. Yet what stuck in the mind was the scar from the Second World War evident in an exchange between Adzhubei and Kennedy on the subject of Germany:

Adzhubei: "In our country there is not a single family that did not lose some kin in the war. You know we are trying to put out the smoldering coals of the last war in Central Europe. But we do not wish only to play the role of political fireman, as it were, though it is very important. In the heart of every Soviet citizen, there are, as you know, coals still burning from the last war and they are burning his soul and do not let him sleep quietly."

Kennedy: "Let me say that I know that the Soviet Union suffered more from World War II than any country. It represented a terrible blow, and the casualties affected every family, including many of the families of those now in government.

"I will say that the United States also suffered, although not so heavily as the Soviet Union quite obviously. My brother was killed in Europe. My sister's husband was killed in Europe . . ."[1]

Another reminder of the traumatic effects of World War II, extremely ordinary and yet extraordinary, came in a photograph released a few years ago by the Novosti Press Agency, a Soviet feature and picture bureau. The

photograph showed a group of Russian women on a
snow-covered plain the morning after a Nazi massacre
of a neighboring village. Scattered across the bleak landscape beneath an overcast sky are scores of corpses
twisted and frozen in the snow. The expressions of grief
and anguish on the women's faces leave an indelible
impression.

These two examples, among others, illustrate why it
is impossible to live in the Soviet Union and not be
continually aware of the extraordinary shock caused by
the surprise attack of June 21–22, 1941. The Soviet
leadership makes a special effort to keep these memories
alive for a younger generation which no longer has any
direct recollection of the war. Pictures, articles, statements, films remind the Soviet citizen that thousands of
farms and factories in the western part of the country
were destroyed by the Nazis; that twenty million people
perished; that the total recorded material damage
amounted to 679 billion prewar rubles; that the nation's
wealth was reduced by roughly 40 per cent.[2]

The shock suffered by Soviet leaders and military
men was profound and is not forgotten. The Soviet
Union had concluded a non-aggression pact with Hitler's
Germany in 1939 and, despite warnings from spies and
allies in 1941, Stalin did not appear to believe that the
Nazis were about to invade. Yet the invasion came—
in the dead of night. The lesson of this experience has
become a tenet of contemporary Soviet military faith;
such an attack, such a war must never be allowed to
happen again. The Soviet Union must never again be
caught off-guard or ill-defended.

The Second World War brought the Soviet Union,
United States, Great Britain, and France into a wartime

alliance of necessity, an alliance which sowed hope for the future despite a backdrop of discord. It brought hope because of common sacrifice, common support, and reasonably good understanding between President Roosevelt, Prime Minister Churchill, and Generalissimo Stalin. But the alliance also contained inherent disharmony because of the ideological tensions between Communism and Western capitalism which went back to 1917 and before. Churchill told Stalin at the Yalta conference, February 1945, that dissension had not been a problem among the wartime Big Three leaders, but that it would plague their successors.[3] By the end of the Potsdam meeting in August, Stalin's partners were no longer Roosevelt and Churchill. Roosevelt had died in April 1945, and Churchill was voted from office in the middle of this conference which had convened to sort out the future of Germany. Discord did, indeed, plague the new Big Three, President Harry Truman, Prime Minister Clement Attlee, and Stalin.

With hindsight, the sources of discord during the later war years are not difficult to understand from the Soviet point of view. Stalin wondered why the Western allies failed to open a promised second front in Europe in 1942. He expressed displeasure over the refusal of the United States to supply more lend-lease aid at a time when Soviet forces were making enormous sacrifices on the Eastern front. On the other hand, American and British leaders saw evil motives in the obstacles which the Russians raised against the use of Soviet bases by Allied aircraft for "shuttle bombing" raids on Germany. And the West was alarmed at Stalin's devious machinations to secure pro-Communist regimes in Rumania and Poland.

Both at Yalta and Potsdam the Big Three had different approaches to the subject of German reparations, and these differences were ultimately important to Soviet successes in space. The Big Three agreed that the Soviet Union had suffered the most from Nazi aggression and should be recompensed the most generously. There agreement stopped. The United States and Great Britain tried to work out a regularized system for extracting reparations from postwar Germany; the Russians talked in terms of punitive measures involving a drastic dismantling of German industry and payment of twenty billion dollars in compensation. Discord at Yalta was settled at Potsdam by agreeing on occupation zones of Germany, and acceding to the Soviet proposal that each occupation power should find reparations in its zone in the form of "war booty."[4] This system, administered by the Allied military authorities in the occupied zones, often amounted in the Soviet case to plunder and deportation of labor.

It was against a background of ideological conflict, the trauma of an initial military disaster, and an uncertain future with capitalist allies that Stalin approached the problem of reconstructing the Soviet Union in 1945 and 1946. On February 9, 1946, he addressed an audience at the Bolshoi Theater before the March elections to the Supreme Soviet (the Soviet parliamentary assembly), with a historically significant speech:

"What are the results of war?

"Our victory means, first of all, that our Soviet social order has triumphed . . . Second, our victory means that our Soviet state system has triumphed. Third, our victory means that the Soviet armed forces have triumphed . . .

"Now a few words about the Communist Party's

plan for the work in the immediate future . . . The principal aims of the new [1946–1950] Five-Year Plan are to rehabilitate the ravished areas of the country, to restore to the pre-war level industry and agriculture, and then to surpass this level . . . Not only to overtake but to surpass the achievements of science beyond the boundaries of our country.

"As regards plans for a longer period ahead, the Party plans to organize a new and mighty upsurge in the national economy, which would allow us to increase our industrial production, for example, three times over as compared with the pre-war period . . . Only under such conditions can we consider that our homeland will be guaranteed against all possible accidents."[5]

The reference to "all possible accidents" was clearly an allusion to the accident of 1941—the Nazi invasion —and the lesson which flowed from it: the Soviet Union must never again be caught off-guard or ill-defended. In 1945, the Soviet Union, like the United States, had emerged as a superpower. And yet the strategic fact in 1945 was that only the United States possessed the atomic bomb and a long-range air force to deliver it. It was inevitable that under such circumstances a centralized governing apparatus, such as Stalin commanded, should address itself to overcoming the "atom gap" and "the delivery gap" as quickly as possible and without much publicity. According to observers, Stalin appeared almost uninterested at Potsdam in 1945 when President Truman disclosed to him that the United States possessed the atomic bomb. It was obviously a feigned lack of interest.

Before the war, Soviet scientists had done sophisticated research into the structure of the atom before their

work was interrupted by hostilities. There is evidence
from Soviet sources that, by 1944, work was resumed
on the development of the atomic bomb. During 1946,
Soviet leaders, including Stalin, denounced the American nuclear monopoly as a menace and promised that
it would not last long. In 1946 also, the Soviet Union
rejected the American Bernard Baruch's proposal to
place all atomic energy under international control in
order to assure its peaceful uses. By the thirtieth anniversary of the Russian Revolution Foreign Minister
Vyacheslav I. Molotov was able to announce that the
atomic secret had been cracked. Two years later, President Truman announced that the United States had detected the first Soviet nuclear test. On September 25,
1949—two days after Truman's statement—the Soviet
news agency TASS confirmed the news that the Soviet
Union had triggered a successful atomic explosion on
August 29, 1949. Vasily S. Yemelyanov, a former chairman of the Soviet Atomic Energy Committee, wrote a
memoir of academician Igor V. Kurchatov, "the father
of the Soviet atom bomb," in which he recalled the intense atmosphere just before the first Soviet nuclear test:

"Indeed: five years of work; billions of investment;
many thousands of persons occupied with atomic work;
all scientists worked up; the whole country on edge . . .

"One can't forget about the international situation;
it was 1949, the height of the cold war. If you take
American newspapers of that time, you would find more
than a tenth of the articles contained threats to the
Soviet Union; calls of various American senators to drop
an atomic bomb on the U.S.S.R. while the Soviet Union
still does not possess it.[6]

"The atomic bomb was for us a necessity. I am deeply

convinced: if there were no danger we would have relaxed, slept; none of us ever would have been occupied
with it.

"We understood perfectly well: we need the bomb so
that we can get on with the work which we have been
conducting since 1917. Without the bomb, we would
have been deprived of the possibility of quietly proceeding with this work. For that reason it was necessary to
pour all our energies into the solution of atomic problems; create the bomb, and continue peaceful construction."[7]

Such was the fervor that went into making the Soviet
atomic bomb. Equal intensity went into solving the
problem of how to dispatch it to its target. Work in this
field proceeded along two parallel courses: development
of the long-range bomber and of rocket vehicles.

In rocket development, Soviet attention naturally
focused on Germany in the immediate postwar period.
It was, after all, the Luftwaffe which had developed the
V1 pulse jet, and the German Army had used the V2
ballistic missile to bombard Great Britain in 1944. The
Red Army entered the German rocket development
center at Peenemunde on May 4, 1945, and seized other
centers in the Harz Mountains where valuable equipment
had been hidden. A year earlier, the Soviet government
had begun preparations for dismantling German industry, and as the Red Army advanced, Soviet teams followed in a relentless search for anything that might be
useful. The Soviet Union's harsh and sometimes punitive attitude had counterproductive effects. Cooperation
with the remaining German experts was frequently difficult; trust was often lacking; and gifted specialists such
as Dr. Wernher von Braun, ultimately a top scientific

figure in the U.S. postwar space program, fled to surrender to the advancing American forces.

Between 1945 and 1946, the Kremlin appeared still undecided on the exact use of German talent. There is evidence in accounts contributed by former Soviet occupation officials of a brief debate over the benefits of putting German industry to work in Germany for Soviet aims, or dismantling whole factories and deporting thousands of Germans to the Soviet Union to work there. At first, special construction bureaus (OKB) were established in factories to study advanced German technology, but finally the idea of dismantling was adopted. Thus, on the night of October 21–22, 1946, about forty thousand Germans were rounded up, coerced into agreeing to five-year contracts, and transported to the Soviet Union.[8]

Meanwhile, the Kremlin leadership had decided to examine the Nazi V2 in detail and attempt to improve it. The V2, which represented a considerable advance over the Russian rocketry of the 1930s, still had shortcomings from the Soviet point of view. For one thing, it was not maneuverable, but traveled along a fixed course; for another, the V2 rocket engine, which developed what was then an extraordinary twenty-five tons of thrust, was only capable of firing the weapon 250 kilometers. Clearly, for Soviet strategic purposes, the V2 was only a start.

Accordingly, the Soviet government began reorganizing its research into rocket and jet propulsion, which, before the war had been conducted by the Jet Scientific Research Institute. By decision of the Council of Ministers, an Academy of Artillery Sciences (later the Academy of Rocket and Artillery Sciences) was established in 1946 under the Defense Ministry. Its mission was to

develop military rockets as well as to train rocket specialists and commanders. Lieutenant General Anatoly
A. Blagonravov, a military ballistics and artillery specialist of long experience, was named to head the new
academy. Blagonravov in a subsequent reincarnation became a prominent spokesman and manager for the
Soviet space program.[9]

There is a great shortage of published documentation
on decision-making during the late Stalin years of 1946–
53, especially on the development of the strategic weapon, because of its classified nature. Some general information about rocket development during this period has
come from the two thousand-odd German rocket specialists who were shipped to Moscow. Many returned
to the West at the expiration of their contracts, but
their reports were inadequate for gaining an over-all
view of the Russian program. Helmut Grottrup, one of
the most prominent captured German scientists, and
others have reported that the Raketenkollektiv—the
"rocket collective"—was scattered about and strictly
isolated from the Soviet mainstream. The Russians, according to this German guidance specialist, distributed
problems to the foreign scientists, forced them into
competition with each other, and with Soviet rocket
groups. Emphasis was on extreme technical simplicity,
partly because of the difficulties of the postwar economic situation in the Soviet Union and partly in reaction to the complexities of the V2 technology evolved
at Peenemunde. The Russians in their search for enormous rocket power reportedly sought to eliminate every
superfluous soldered part and joint; they relied on ordinary metals; they did not spend much time in the laboratory experimenting with exotic fuels, but preferred to
depend, instead, on liquid oxygen and kerosene.[10]

Generally speaking, the first objective was to produce a Soviet V2 with twenty-five tons of thrust. After this was done, Soviet scientists in the late 1940s augmented the V2 engine with from twenty-five to thirty-five tons of thrust. The United States did not match this power with the Redstone rocket until 1951 under the direction of Dr. Wernher von Braun. The aim of the Soviet's 1946–50 Five-Year Plan in rocketry was to produce a reliable rocket weapon, and Soviet scientists managed to do it. The first military divisions equipped with "Pobeda" (Victory) rockets were reportedly organized in 1950–1. As to the contribution of the conscripted German scientists, it had been mostly complementary. At best, Grottrup estimated, the Germans showed that it would be technically feasible to produce a single-stage, medium-range rocket. Such a vehicle, if successfully developed, could hit a target 1,800 miles away.[11]

It was at this point that the United States, Britain, and France began to notice that the Russians were making important strides in rocketry. One of their first sources was Colonel Grigory A. Tokaty-Tokaev, who had served since 1945 as the Soviet government's chief rocket specialist in Germany. He fled to Great Britain after two and a half years of duty with the Soviet Military Administration in Germany. Twenty years later in London, where he was the head of the Department of Aeronautics and Space Technology at the City University, Tokaty-Tokaev recalled in some detail the motives of the Soviet government's rocket program in the postwar years: *

"You see, the U.S.S.R. claims to be a Communist country; and the long-term aim of Communism is the re-

*For the full text of the author's interview with Tokaev, see Appendix A, page 219.

placement of the capitalist system by the system of Communism. Now, Marx, Lenin, Trotsky and Stalin taught that this aim can only be achieved by means of a socio-political and economic revolution. But such a revolution requires modern armaments. Moreover, the theoretical aim of Hitler's war against the U.S.S.R. was the destruction of the achievements of the October Revolution and, consequently, the prevention of such revolutions elsewhere. But the Soviet Union came out of World War II as a leading military power, determined to stand up to any new anti-Communist war, to the whole non-Communist world. Above all, this meant standing up to the United States, which by then possessed the B-29 bomber, the A-bomb, many German V2s, and the leading rocket designers of Germany. The Soviet leaders knew that until and unless they did something along these lines, they could not stand up to the U.S.A. To the Russians there was nothing new in rocketry, in general, and in a theoretical sense. But during the war the U.S.S.R. had not produced anything like the V2 rocket; the Kremlin leaders were very worried . . . In rocketry proper, too, there had been good progress in the Soviet Union. But then there was war. In the beginning our armed forces were smashed. We had to dismantle everything and move back to a safe area. We lost every normal condition of work. The war abruptly distorted our work. The U.S.S.R. was losing everything, and the Germans were gaining a great deal from other Europeans. We could do little until the end of the war. We had to move back and then start again at square one. In the end, however, we learned a great deal from the Germans, other European countries, from British and American experience. We were anxious to learn from everyone, which helped.

"By the end of World War II, the Soviet Union and

the United States constituted two profoundly different worlds. Now, remember that the U.S.A. had the long-range B-29 bomber and the A-bomb: we could be reached by them. But we had neither a bomber capable of reaching the U.S.A. nor the A-bomb. From a purely military point of view, the situation was really desperate, and hence the line of thought of the Kremlin leaders. There was only one way out: to solve the problems of long-range bombers, rockets, and A- and H-bombs.

"This was a direct confrontation of the old enemies—of Communism and Capitalism. America was the determined leader of the second, and the Soviet Union was the determined leader of the first. The so-called 'proletarian internationalism' made the U.S.S.R. responsible for Communism at large; not just for the U.S.S.R. itself, but throughout the world. For all theoretical and propaganda purposes; this was a great responsibility before the history of Communism. In reality, however, it was a responsibility before the age-long tradition of Russian expansionism. But whatever it was, it required, or demanded, the urgent creation of appropriate material means. What were these means? Well, we began working on an aircraft similar to the B-29. Already in 1944, Soviet scientists knew of the V2 and the Sanger project. And it was natural for scientific advisers to call the attention of the leaders to these projects: governments do not make decisions without consultations with advisers."[12]

The Sanger project, to which Tokaty-Tokaev referred, was a proposal to the Germans by an Austrian scientist, Eugen Sanger, for an intercontinental bomber, boosted by rocket power and capable of traveling at high altitude and great speed. The project interested Stalin.

Tokaty-Tokaev was dispatched to Berlin in June 1945

from the Zhukovsky Military Air Academy in Moscow, where he had occupied the position of lecturer in the aerodynamics of flight. He had also served simultaneously as professor of aviation at the Moscow Engineering Institute. During the following months in Germany he was the jet propulsion expert of the Soviet Military Administration under Marshal Georgy K. Zhukov, supreme military commander of the Soviet Zone. It fell to Tokaty-Tokaev to make an assessment of German achievements in rocketry. On examination of the remains of Peenemunde and other German rocket centers, Tokaty-Tokaev concluded that the Nazis had surpassed the Russians in rocket technology, but that Russian rocket *theory* was every bit as advanced. This was obviously an enormous tribute to the prewar work of the rocket scientists and theoreticians employed by the Jet Scientific Research Institute.[13]

In a taped interview for the National Air and Space Museum of the Smithsonian Institution in Washington, D.C., Tokaty-Tokaev discussed some of the difficulties he encountered in convincing higher Soviet authorities in 1946–7 that Soviet scientists were capable of producing a rocket booster and an artificial satellite without dependence on the Germans. He ran into interference from Stalin's son, Major General Vasily I. Stalin of the Soviet Air Force, and also from General Ivan A. Serov, deputy supreme commander of the Soviet Military Administration in Germany and a first deputy of the Soviet state security apparatus. In Germany, Serov played a crucial role in the search for industrial secrets and for personnel who could aid the development of the atom-bomb project as well as rocket research.

By 1947, less than a year after the removal of the

a transoceanic rocket bomber as both a strategic weapon and potential bargaining tool at any negotations with his prime adversary, President Truman. This meeting also heard recrimination for the Soviet failure to capture German specialists of the highest caliber. Stalin declared: "It seems to me a very poor sum total, if we smashed the Nazis, took Berlin, took Vienna, but the Americans got Von Braun, Lippisch; the British, I am told, got Buzemann, and perhaps Tank, and now the French have got Dr. Sanger."[16]

The question next arose about how to proceed with building a long-range rocket weapon. Stalin inclined toward exploiting the German specialists and, further- more, to kidnapping more of them wherever possible. Finally, a decision on forming a high-powered rocket development committee was approved. By a secret Council of Ministers decree, April 17, 1947, the Gov- ernment Commission for Long-Range Rockets (Pravitel- stvennaya Kommissiya po Raketam Dalnego Deistviya) was established under chairmanship of Colonel General Serov, with Colonel Grigory A. Tokaty-Tokaev as dep- uty chairman, and Mstislav V. Keldysh of the Arma- ments Ministry (subsequently a very prominent space official and scientist as President of the Academy of Sciences, beginning in 1961), M. A. Kishkin of the Ministry of Aircraft Production, and Major General Vasily I. Stalin. The commission's assignment called for a feasibility study to be presented to the Council of Ministers by August 1, 1947. The commission was also to go to Germany, scrutinize German contributions and experts, and contemplate the selection of personnel to work on the Sanger and other projects.[17]

Several paradoxes emerged out of this period. First,

members of the top Communist Party leadership were alerted to the possibility of using powerful rocket boosters for orbiting a satellite and pursuing the exploration of near-space. This proposal, which fascinated a number of prominent Soviet scientists, did not capture the imagination of the Communist Party's top echelon. Nor, apparently, was it even seriously discussed. Second, the propaganda value of transoceanic rockets, or even of a pioneering satellite with which to stun the world, was only hazily understood. In an interview with the author, Tokaty-Tokaev said:

"I do not remember specifically propaganda statements at these meetings. I thought Stalin and his colleagues meant business. But, then, of course, every stick in the world has two ends: a rocket—Soviet or American—is both an effective monster and a propaganda weapon. I also agree that the Soviets love propaganda. But it would be a dangerous illusion to think that Gagarin, Titov, Bykovsky, Nikolayev, Tereshkova, Popovich, Komarov, Belyaev, Feoktistov, and Yegorov were nothing more than propaganda. Facts are stubborn things that do not cease to exist because they are painted this or that color.

"Having said this, I should now like to focus your attention on something else. You see, the peculiarity of space technology in the U.S.S.R. is that things are designed to fulfill two or more simultaneous functions. Sputnik-1 was a scientific achievement, a heraldic symbol over the gateway into the unknown, a challenge-warning to the capitalist West, and an outstanding propaganda drum, etc. And the designers were aware of all these functions. Similarly, the emergence of the purely strategic ICBM was something like a proclamation of the beginning of space exploration, of man's flight around the moon and beyond, as well as of Sputnik, etc. In other words, there are no rockets, spaceships and space pilots devoid of propaganda value, and there has never been a propaganda launching devoid of scientific technological importance. Therefore, he who talks of rockets and sputniks in terms of only propaganda should have his head examined."[18]

The main thrust of the Kremlin deliberations about Sanger's bomber was the effort to improve the Soviet Union's military power. Whether the effort was viewed as a defensive response or a frankly aggressive move is not entirely clear because of the relative lack of detail. The concern of the Soviet leaders over American nuclear capabilities shows a definite preoccupation with defense. The experience of the recent past and the lessons of 1941 would seem to underline this preoccupation. Yet, Tokaty-Tokaev came away from these meetings with the impression that the ultimate interest of the Soviet leaders was in attack weapons. His impression of the bellicose atmosphere at the April meetings contributed to his decision to flee to Great Britain in 1948. His views on "Kremlin warmongering" were recorded in a letter to *The New York Times* published September 4, 1948, and in a pamphlet entitled "Inside the Kremlin" published in 1949. He finally wrote a short book called *Soviet Imperialism,* published in 1951, which also described the allegedly aggressive intentions behind the Soviet rocket program.

At this point, another strand of development—peaceful scientific development—enters the picture. Soviet scientists had alerted their political leadership to the possibility of using the new, giant boosters for orbiting an artificial satellite and taking probes of the earth's atmosphere of a kind never before performed. Their suggestions, in the late 1940s, however, had fallen on deaf ears. By the early 1950s, the situation was changing. For one thing, scientists had adapted less powerful rockets to take vertical soundings, and such geophysical and meteorological probes were carried out beginning in 1946. From 1949 onward, the Soviet Academy of Sciences had also arranged to make use of the Soviet-built V2s to explore systematically the upper atmosphere, the extent of cosmic radiation, the nature of air currents, and the analysis of the solar spectrum. These experiments were followed, 1949–52, by the launching of fourteen experimental dogs to altitudes of up to four hundred kilometers. The animals were monitored for heart activity, blood pressure, respiration, and body temperature. The exotic conclusion began to emerge: manned rocket flight would eventually be both scientifically and medically possible.

In addition to the Russian scientists' use of rocketry, a movement developed in the West to organize a broad international effort for systematic measurements of the earth's surface, its atmosphere, and its relation to the sun. This movement turned into the International Geophysical Year and eventually provided the Soviet Union with an ideal cover and excuse for supporting its own scientists' ideas about space exploration. The Kremlin, in the end, decided to have another look at the suggestion that an artificial satellite should be launched and that the Soviet Union should consider the possibility of launching men into space behind the shield of the International Geophysical Year.

The first conception of a year for coordinated exploration of the earth's surface and its complicated relations with the sun go back to an evening's discussion among a number of eminent scientists, April 5, 1950. Their proposal, in its original form, was for a follow-up to the First Polar Year of 1882 and the Second Polar Year of 1932. However, as time passed after that meeting (at the home of physicist James Van Allen in Silver Spring, Maryland), the idea shifted toward a more comprehensive year of scientific cooperation. The proposal ultimately envisaged eighteen months of varied geophysical observations from July 1, 1957, to December 31, 1958—a period of intense solar activity. The idea was actively promoted by an international preparatory committee which came to be known as CSAGI after its French name, Comité Spécial pour l'Année Géophysique Internationale.

In May 1952, CSAGI sent out invitations all over the world. The committee's hope was that international politics could be pushed aside in favor of research that would be of ultimate interest to all mankind. An invitation was duly forwarded to the Soviet Academy of Sciences.

May 1954 was set as the deadline for the submission of detailed plans by national IGY committees. This deadline came and went without any word from Moscow as to the Soviet Union's intentions. CSAGI's planning, nevertheless, went forward and the committee convened in Rome for its scheduled session of September 30–October 4, 1954. Suddenly, the Soviet Embassy dispatched a communication. Soviet scientists would participate.[19]

If the Soviet acceptance startled CSAGI, CSAGI jolted the Russians. The Rome meeting adopted a historic resolution that posed before the United States and the Soviet Union an extraordinary scientific challenge: "In view of the great importance of observations, during extended periods of time, of extra-terrestrial radiations and geophysical phenomena in the upper atmosphere, and in view of the advanced state of present rocket techniques, CSAGI recommends that thought be given to the launching of small satellite vehicles, to their scientific instrumentation, and to the new problems associated with satellite experiments, such as power supply, telemetering, and orientation of the vehicle."[20]

Thus, by the last year of Stalin's life, March 6, 1952, to March 5, 1953, an extraordinary confluence of events was beginning. Soviet leaders were confronted with a number of propositions connected by a common thread: the world's scientists wanted to harness contemporary rocket power for eighteen months of international, geophysical soundings; the Soviet military wanted an intercontinental ballistic rocket; Soviet military planners foresaw the potential value of an orbiting satellite as a moving platform for scooping up intelligence information or launching bombs; Soviet scientists realized that satellites could be used for a variety of purposes as well as a stepping stone into space. The year 1954 and the months preceding it now appear in historical perspective to have been a crucial turning point in the history of the Soviet space program.

From Soviet sources it is known that in 1954 the Gas Dynamics Laboratory in Leningrad began the final development of the RD-107 and RD-108 rocket engines which were used to power the original intercontinental ballistic missile and the rocket which launched Sputnik. These engines, as already noted, have continued to power the first stage of the rockets that have launched all the Soviet cosmonauts into space through 1971— a remarkable long-term achievement. Furthermore, it was in 1954 that a space flight commission was established within the Academy of Sciences to coordinate scientific research in connection with the satellite effort. Officially, this group was known by the jaw-breaking title of the Interdepartmental Commission for the Coordination and Control of Scientific-Theoretical Work in the Field of Organization and Accomplishment of Interplanetary Communications of the Astronomical Council of the Agency of Sciences of the U.S.S.R. It was composed of at least twenty-seven distinguished scientists. Among them there were such personalities as academician Pyotr L. Kapitsa, the Soviet physicist who had worked before the war at Cambridge University in England and was to become something of a science adviser to Premier Nikita S. Khrushchev; academician Anatoly A. Blagonravov, first president of the Academy of Artillery Sciences and a prominent military engineering expert; academician Nikolai N. Bogolyubov, the Russian mathematical genius; academician Viktor A. Ambartsumyan, the celebrated Armenian astronomer; Dr. Yuri A. Pobedonostsev, a prewar MosGIRD rocket pioneer; Dr. V. F. Bolkhovitinov, a military scientist attached to the Zhukovsky Military Air Academy and designer of a rocket plane dating back to 1942; and Dr. Georgy L. Pokrovsky, a military explosives expert. The chairman of the commission was academician Leonid I. Sedov, a celebrated gas dynamicist who had been associated with the development of aircraft propulsion.

The announcement of the space flight commission

was given only slight publicity, and not until 1955. On April 16, 1955, there appeared an article in the Moscow evening newspaper *Vechernaya Moskva* that outlined the terms of reference of the commission and gave a list of some of the scientists working on it. The announcement made no secret of the fact that the Soviet Union intended to organize the development of an artificial satellite as a first step toward the further exploration of the universe.[21] There were, of course, many details still unclear about the commission's role but an important fact emerged from an examination of the membership of the group: the Soviet leadership had decided to bring together in a cooperative effort both civilian and military scientists. To save effort and money, the Soviet Union, unlike the United States, rationalized the scientific and military establishments at an early date in its efforts to launch a first artificial satellite.

The launching of Sputnik on October 4, 1957, unleashed "The Storming of the Cosmos." It has frequently been assumed that Sputnik came as a great surprise, that there was no forewarning. This is wrong. There was forewarning, but the great masses of the Western public did not notice it. Soviet specialists in the West had picked up the hints as had Western scientists.

The telltale signs cames in different ways. Some hints were accessible to the general public; some were not. The United States, for example, began systematic monitoring of Soviet airwaves and internal transmissions in the 1950s. From this eavesdropping, it was learned that Soviet scientists were progressing in their development of powerful rocket boosters. By 1955, the Central Intelligence Agency in Washington presented evidence to the National Security Council for consideration by the highest officials of the Eisenhower Administration: the Soviet Union might well be trying to achieve an orbiting artificial satellite before the United States and thereby reap a rich scientific and propaganda victory. The CIA furthermore proposed that the United States should make a concerted effort to foil this thrust.[22]

There were other hints, too, available to the interested outsider. At first the hints, like some obscure sign language, were intelligible only to those who knew the code: a basic knowledge of developments in Soviet rocketry. It would have been known only by a limited number of Soviet officials. Mikhail K. Tikhonravov, who had been a prominent member of MosGIRD and was currently a member of the Academy of Artillery Sciences, wrote in the children's newspaper *Pioneer Pravda* on October 2, 1951, that the first manned rocket flights in the Soviet Union would occur in the period 1961–6. Yuri A. Gagarin, who actually flew in 1961, could have been among the young teenagers who read the newspaper.

At the Communist-sponsored World Peace Council in Vienna in November 1953, the President of the Soviet Academy of Sciences, Alexander N. Nesmeyanov, made a statement which, in retrospect, appears to carry more meaning than might have been supposed at the time. "Science," he said, "has reached a state when it is realistic to launch a stratoplane to the moon; to create artificial satellites of the earth; where there have been found effective treatments for diseases—the most terrible bane of humanity—when the problems of energy have attained entirely new horizons." In 1954, the July issue of *Youngster's Mechanics* (*Tekhnika Molodyezhi*) printed a general space flight timetable. Probably this timetable was an approximate one and did not reflect exact Soviet intentions. But it did symbolize the growing interest in putting man in space. The timetable predicted the launching of an artificial satellite in 1965; launching of a spaceship carrying three men in 1975; circumnavigation of the moon in the period 1980–90; and a manned landing on the moon by the year 2000.

As time went on, the signals became more frequent and clearer. Soviet leaders talked occasionally in general terms about their progress in rocketry, and *Pravda* spoke frankly about "America's lag."²³ On July 29, 1955, the White House announced that the United States would attempt to orbit a number of small scientific satellites during the International Geophysical Year. Three days later, on August 2, 1955, academician Leonid I. Sedov, the chairman of the Interdepartmental Commission on Interplanetary Communications, called a press conference in Copenhagen, where he was attending the International Astronautical Federation Conference. There he announced similar Soviet plans for a satellite. He noted that technologically it would be possible for the Soviet Union to launch a satellite within the next two years and broadly hinted that the probe could be heavier than what the United States had under consideration. Characteristically, he declined to give a specific date for the launch.

In the year of Sputnik—1957—other signals flashed. President of the Academy of Sciences Nesmeyanov announced June 9, 1957, that Soviet scientists had theoretically solved the problem of orbiting an artificial satellite. The academy's *Astronomical Journal* announced that Sputnik would broadcast on frequencies of 20 and 40 megacycles. Then on August 26, 1957, came a stunning announcement that could not help but force Western military planners and less formal observers to take stock. The news agency TASS announced that the Soviet Union had tested "a few days ago" super long-range intercontinental rockets over its territory. Actually these tests were carried out in the months of July and August, and there are indications that the scientists were satisfied with their achievements following a conclusive test August 3, 1957.*

The announcement was sensational because the United States had not yet carried out similar long-range tests in its own ICBM program. Military experts were quick to grasp the strategic implications of the TASS statement—assuming, of course, the Russians were well on the way to a capacity for hitting the United States with nuclear missiles. TASS was instructed to use the occasion to drive home a favorite theme of Soviet public policy: The U.S.S.R. really seeks peace, but the Western countries have failed to take practical steps towards disarmament. Consequently, the Soviet Union has been forced to take all necessary measures to defend itself, although this does not mean that in the future it would abandon the search for peace and disarmament.

August 26, 1957, also signaled a most serious possibility that Soviet scientists would be capable of orbiting an artificial satellite during the International Geophysical Year. Speculation began to grow in the West that the launch was, indeed, imminent. There were those

*Soviet sources avoid giving the exact dates of the ICBM tests, but Moscow Radio has indicated that August 3, 1957, was the date of one of the successful tests. In a home service broadcast on August 2, 1967, Moscow Radio said: "Tomorrow will be the tenth anniversary of the successful test in the Soviet Union of the first ICBM in the world."

who anticipated it on September 15, 1957—the one hundredth anniversary of Konstantin Tsiolkovsky's birth. The anniversary came and went but there was no launch. The launch came two weeks later. Sputnik was fired at a time when an IGY committee was meeting in Washington to coordinate rocket and satellite launchings during the International Geophysical Year. Anatoly A. Blagonravov, head of the Soviet delegation to the IGY committee meeting, was all smiles.

There are three interesting vignettes of the efforts to orbit *"prostyeshi sputnik"* (the simplest satellite) as it was called by the engineers who worked on it. These views show with what enthusiasm Soviet engineers approached the event; with what clinical detail and political overtones TASS made the announcement to the world; and how offhandedly Nikita S. Khrushchev, the Soviet Communist chief, greeted the news.

Aleksei Ivanov, an engineer who worked in the special Sputnik bureau, recalled that it was August when preparations for the launch went into high gear. Weeks of intense work began in early September, and occasional quarrels flared up during the prelaunch tension. At one point, as the launching date drew near, an embarrassing failure in Sputnik's electrical system occurred which threw everybody into a momentary panic. Finally, on the morning of October 3, the countdown began. The hours ticked by. With less than half an hour to go, on October 4 the technicians retired to their observation bunker. The words are Ivanov's:

"Here is the observation point—it is about one kilometer from the pad. We clamber onto the roof of the radio station to see better. The hands of the clock get closer to the designated minute. What a tiringly long

time it is taking! Now, here . . . here it is now. . . . My
heart seems to be bursting in my chest. Why so long?
What long, drawn-out seconds!

"I watch, not moving my eyes away, fearing I might
blink, so as not to miss the moment of liftoff.

"At last the flash of flames and on its heels the roar.
A low, rumbling roar. The rocket is shrouded in flames;
the fumes come on thick and fast. It seems that suddenly
they will completely cover the rocket, but finally, like a
triumphant victory hymn, the white body of the rocket
begins moving, takes off . . . and the flash, the brightest
flash of light! The beam rips through the dark of night.
I look down on the ground. It's bright, very bright.
There are shadows; sharp, creeping, black shadows
from the people and equipment.

"The rocket is flying. Ever faster and faster. Higher
and higher. Now the smooth tip-over into its trajectory;
it seems like the flames are beating straight into my
eyes. But the distance softens the light. The roar becomes duller. The contours of the mighty body are no
longer visible. Only the flash is still clear, but with each
second it grows dimmer. Finally, it is only a tiny star
disappearing into the night. You can't recognize it
among the many present stars.

"A moment's silence. . . . And . . . a shout. Everyone
is shouting. What they are shouting, you couldn't figure
out; they are waving their arms, embracing each other,
kissing each other. Someone is rubbing his unshaven,
tickly beard into my cheek; someone is whomping me
on the back. Happy faces and but a single word . . .
'She flew!'

"We realize after a few minutes that all has begun
well. The operators at the receiving station report that

all is within the norm. They have received signals of the ejection of the satellite from the rocket; the necessary velocity has been achieved. There it is. Orbital velocity!

"An hour goes by. Impatiently we drag ourselves to the receiving station where you could hear the radio signals of the satellite.

"The seconds drag on tiresomely. In the station it's as quiet as the grave. All of a sudden Vyacheslav Ivanovich raises his head; adjusts his earphones; adjusts the set and, without raising his eyes, announces loudly, but without certainty:

" 'It seems all right . . .'

"Just a few seconds more (each second the satellite gets closer to the speed of eight kilometers a second). And there she goes!!

"The earphones come off his head. You can already hear even more clearly the proud, familiar: Beep-beep! Beep-beep! She's flying! She's flying!"[24]

In the early hours of Saturday, October 5, 1957, an employee of the Soviet news agency TASS bent over a teletype transmitting machine in Moscow, inserted a punched tape, and flicked a switch. From the TASS headquarters on the edge of the tree-lined Tveroski Boulevard, a message began to clatter out to the world:

For several years research and experimental design work has been underway in the Soviet Union to create artificial satellites of the earth. It has already been reported in the press that the launching of earth satellites in the U.S.S.R. was planned in accordance with the International Geophysical Year.

As a result of the intensive work by research institutes and designing bureaus, the first satellite was successfully launched in the U.S.S.R., October 4.

According to preliminary information, the carrier rocket has imparted to the satellite the required orbital velocity of 8,000 meters [about 26,000 feet] per second. At the present time the satellite is describing elliptical trajectories around the earth. Its flight will be observed in the rays of the rising and setting sun with the aid of the simplest optical instruments and spy glasses.

According to calculations which are being supplemented by direct observations, the satellite will travel at altitudes of up to 900 kilometers [about 640 miles] above the surface of the earth. A complete revolution of the satellite will take one hour and 35 minutes. Its orbit is inclined at an angle of 65 degrees to the equatorial plane. Tomorrow the satellite will pass twice over the Moscow area at 1:46 P.M. and at 6:42 A.M. Moscow time.

Reports about the subsequent movement of the first artificial satellite launched in the U.S.S.R. on the 4th of October will be issued regularly by Soviet broadcasting stations.

The satellite is of spherical shape, 58 centimeters in diameter and weighing 83.6 kilograms. It is fitted with radio transmitters continuously emitting signals at a frequency of 20.0005 and 40.002 megacycles or 15 and 7.5 meter wave lengths respectively.

The power of the transmitter is such as to ensure reliable reception by a broad range of amateurs. The signals are of the nature of telegraphic signals at about 0.3 seconds duration. The signals of one frequency are sent during pauses in the signals of the other frequency.

Scientific stations at various points in the U.S.S.R. are conducting observations of the satellite and determining elements of its trajectory. Since the density of the rarefied upper layers of the atmosphere is not accurately known there are no data available at present for determining the exact period of the satellite's existence or the point of its entry into the denser layers of the atmosphere.

Calculations have shown that owing to the tremendous velocity of the satellite at the end of its existence, it will burn up on reaching the denser layers of the atmosphere at an altitude of several hundred kilometers.

The possibility of cosmic flight with the help of rockets was first scientifically substantiated in Russia as early as the end of the 19th century in the works of the outstanding Russian scientist Konstantin Eduardovich Tsiolkovsky.

The successful launching of the first man-made earth satellite makes a tremendous contribution to the treasure house of world science and culture. The scientific experiment staged at such a great height is of great importance for fathoming the properties of cosmic space and for studying the earth as part of our solar system.

The Soviet Union proposes to send up several more artificial satellites during the International Geophysical Year. They will be bigger and heavier satellites and will help to carry out an extensive program of scientific research.

Artificial earth satellites will pave the way for space travel and it seems that the present generation will witness how the freed and conscious labor of the people of the new Socialist society turns even the most daring man's dreams into reality.[25]

On October 8, 1957, four days after Sputnik went into orbit, Communist Party First Secretary Nikita S. Khrushchev received James Reston of *The New York Times* at the Party's Central Committee headquarters in Moscow. During the ensuing three and a half hours, the two men talked about a multitude of subjects, military, political, scientific. Khrushchev said that he had not attended the launch. In fact, he had been vacationing in the Crimea and was on his way back to Moscow when Sputnik went up. Reston observed that Khrushchev's attitude on the launching of Sputnik, which caused such an enormous sensation in the West, was "almost casual." In their conversation, reported in the transcript published by both *Pravda* and *The New York Times,* Khrushchev said simply, "When the satellite was launched, they phoned me that the rocket had taken the right course, and that the satellite was already revolving around the earth. I congratulated the entire group of engineers and technicians on this outstanding achievement and calmly went to bed."[26]

Oddly enough, the Soviet leaders did not seem to have foreseen the full political impact of the launching of Sputnik. It would have been difficult to predict with accuracy the uproar which was to follow, particularly in the United States, where almost every Congressman and political leader issued every conceivable comment. The Kremlin leadership probably realized that Sputnik would boost Soviet prestige, and there is evidence which will appear in a later chapter that Khrushchev and his colleagues made a conscious effort to beat the United States into space. Yet the First Secretary of the Communist Party went to bed calmly on the night of October 4, 1957, and it was not until several weeks later that he began to capitalize on the drama of the new era.

chapter 3

SECRET
SCIENTISTS

In November 1963, foreign correspondents stationed in Moscow came, unexpectedly, face to face with a Soviet state secret. There was, they were told, one Soviet scientist who was unusually influential in planning and directing the Soviet space program. His name was Sergei Pavlovich Korolyov. Furthermore, another very important person in the Russian space effort was Valentin P. Glushko.

The correspondents immediately began to check out these reports. But secrets are secrets; they are not subject to easy verification by curious foreigners. Soviet officialdom professed ignorance of Korolyov and Glushko. Soviet journalists, an occasional source of accurate unpublished information, were vague. Soviet reference books recorded the existence of the two men but a total blank surrounded their current occupations. Nevertheless, the manner in which the report had come to the correspondents obliged them to give it some credence: it emanated from the famous wedding reception that Khrushchev gave in Moscow on November 3, 1963, for Valentina V. Tereshkova, the first woman in space, and her bridegroom, Andrian G. Nikolayev, in the Government Reception House. Western newsmen were invited to the wedding ceremony and to the reception with a panoply of important Soviet space officials. Among them were Korolyov, Glushko, and others whom the correspondents could not recognize by face although they learned through informal conversation of their presence among the crush of guests.

What the correspondents did not know was that Korolyov and Glushko had been identified as important contributors to Soviet rocket development in September 1961 at a meeting of the British Interplanetary Society in London. The speaker on that occasion had been Tokaty-Tokaev, who gave one of the earliest authoritative outlines of the history of rocket development in the Soviet Union now generally available in the West. Soviet diplomats in the audience presumably reported these revelations back to Moscow, while the main contents of Tokaty-Tokaev's lecture gained only limited circulation.[1]

Also, unknown to correspondents in Moscow, U.S. government analysts were coming independently to the conclusion that Korolyov was an important figure, most likely the mysterious Chief Designer. The United States' study was based on extensive research of open Soviet

literature between 1930 and 1964, which showed that
Korolyov and Glushko, and Mikhail K. Tikhonravov,
were probably leading figures in the Soviet space program.[2] The analysis disclosed that these three, along
with others, had worked in rocketry before the Second
World War. Their contributions were described in
books and periodicals of that era, but as the space
program progressed into the mid-1950s references to
them disappeared from Soviet publications. By 1958
and 1959 censorship caused any description of their
contributions to be severely edited. The result was that
their existence was acknowledged, but no one and no
publication would say what these men were doing.

In the end, some correspondents obtained from the
TASS photo archives a file picture of Korolyov and of
Glushko. These were sent abroad with cautious captions
indicating that the two scientists might be prominent
space officials. *The New York Times* was bolder and
more perceptive. Correspondent Theodore Shabad
wrote in the November 12, 1963, issue:

MOSCOW, Nov. 11—Reports circulating in Moscow's
Western community last week have mentioned two rocket
pioneers as likely figures in the Soviet space program.

Although the identities of the top scientists in their jobs
remains an official secret, a number of official reports have
been pointing to two academicians, Valentin P. Glushko,
a combustion engineer, and Sergei P. Korolyov, a mechanical engineer.

These reports cannot be confirmed from official Soviet
sources. The leading figures in the Soviet space effort have
been cloaked behind such designations as Chief Designer
and Chief Theoretician, which always appear in the Soviet
press with capitalized initials.

According to available biographical data in official refer-

ence books, Messrs. Glushko and Korolyov have had roughly
parallel careers in the last 10 years. They were admitted to
the Soviet Academy of Sciences as corresponding members
in 1953. They were promoted to the rank of full academicians in 1958, one year after the launching of the Soviet
Union's first artificial satellite.

The Academy, which includes the highest ranking scientists of the Soviet Union, is made up of about 160 full
academicians, and about twice as many corresponding
members.

According to the Great Soviet Encyclopedia, Mr. Glushko
participated in the first Soviet experimental work on developing liquid fuel rockets in the early nineteen thirties. This
work was directed by Friderikh Tsander, an early Soviet
rocket engineer of Baltic-German origin who died in 1933.

Working with Mr. Tsander, Mr. Glushko was said to have
designed rockets that used liquid oxygen and nitric acid
as fuel.

Mr. Korolyov seems to have been connected with the mechanical and structural aspects of rocket flight. In 1940, according to the encyclopedia, he designed a glider that was to
be powered by an experimental liquid fuel rocket engine.

Before his election to the academy, Mr. Glushko was
identified officially as a laboratory head in the Solid Fuels
Institute, and Mr. Korolyov as a laboratory head in the
Institute of Machine Studies.

The secrecy surrounding the Soviet program is not
immutable. It unravels with the passage of time, with
the obsolescence of equipment, with the death of
prominent individuals, with little hints and leaks that
surface at unexpected and unlikely moments. Slightly
more than two years later, Korolyov was dead. He died
on an operating table in a Moscow hospital on January
14, 1966, and two days afterward was at last officially
acknowledged to be the Chief Designer of Rocket-
Cosmic Systems. The veil of secrecy then again de-

scended on the new chief designer, or the group of men who succeeded Korolyov in directing space exploration.

The question of who runs the Soviet space program is a constant and compelling one. The masking of the space leaders' identities introduces an element of mystery that challenges the curious outsider and provides a continuing subject for speculation and informed guessing. More seriously, if the identity of the managers were known, the outside observer would be a little closer to knowing the range of Soviet intentions in space, the priorities assigned to various projects, and most particularly, the value that the Kremlin assigns and did assign, in the 1960s, to a manned lunar landing, which the United States declared it would achieve by the end of that decade.

The struggle to pierce the security veil is annoyingly tough. The authorities who impose and maintain the official secrecy are, obviously, most conservative. They do not like to admit the mere existence of a secret subject, the mere fact that a piece of information is classified. At international conferences Soviet scientists representing the Academy of Sciences' Interdepartmental Commission on Interplanetary Communications have resorted to vague and generalized statements to avoid acknowledging that "x" exists but cannot be discussed because it is secret.

Khrushchev would occasionally depart from these conventions. He was known for his indiscretions, and appreciated because of them by journalists. In 1958, the First Secretary asserted that space secrecy was justified by the demands of military security. This was a time when the Soviet military and the civilian space program were intimately joined in pioneering a new field of technology. Khrushchev promised that as time went by "the photographs and names of these illustrious people will be made public." But for the moment, he added, "in order to ensure the country's security and the lives of these scientists, engineers, technicians, and other specialists, we cannot yet make known their names or publish their photographs."[3]

Thus the facts are scarce; the secret list is long. The Soviet Union has tried to hide the names and locations of its launching sites. It does not announce how much it is spending on space exploration. It talks about the general conquest of space but adheres to the principle of releasing only the most meager details about the mission aims of any particular flight (at least until the flight is successfully completed). It has never published an organizational chart of its space agency and, in fact, generally avoids referring to its space agency—which is not the Academy of Sciences' Interdepartmental Committee. The current director of the Soviet space agency is never publicly named, nor are his colleagues and assistants.

And yet, over the years, a few telling facts have emerged. If one looks at what has been said about the direction of the space program, certain general shapes begin to emerge. In the early days, immediately after Sputnik, Khrushchev used to take credit as the person responsible for directing the Soviet space program. Soviet cosmonauts, traveling abroad on goodwill missions, would stress that it was Khrushchev who was the Soviet space genius. The Communist Party newspaper *Pravda* reported authoritatively in 1961 that Khrushchev visited all factories and test stands, that he knew the leading space scientists by name and face, and that in difficult situations "he participates in the discussions of all the most vital experiments." Khrushchev, said *Pravda,* "directs the development of the major directions of technical progress in the country, and the determination of the basic directions and establishment of generally planned growth of cosmic science and technology. In his able proposals, there is evidence again and again of the great conviction in the triumph of Soviet rocket technology."[4]

There was probably some truth to the assertion that Khrushchev was the director of the space program. As First Secretary of the Communist Party and, simultaneously, Chairman of the Council of Ministers, Khrushchev was in an unusually powerful position. He was able to arrogate to himself the final decision on almost any matter in the country from agriculture to diplomacy or zoology. His interest in space—tentative and hesitant at first—grew ever greater as Soviet successes demonstrated dramatically the triumph of man over nature. The launching of the first human beings into space, Yuri A. Gagarin and Gherman S. Titov, between April and August 1961 created a human drama of the highest order. These space odysseys were played out with Khrushchev's direct involvement and under intense, if carefully controlled, publicity. Khrushchev's responsibility was surely great and he undoubtedly discussed the general directions of "The Storming of the Cosmos." However, one would hardly think of him as the day-to-day manager of a space agency.

The year 1961 did reveal a number of other high officials involved in the space program. Their participation came to light at a Kremlin ceremony, June 20, 1961, at which leading scientists and administrators were decorated for Gagarin's triumphs. Khrushchev was supreme among them. As if to underline his senior role, Khrushchev accepted the highest Soviet decorations for the successful execution of the first manned space flight: the Order of Lenin, a gold Hammer and Sickle medal, and a certificate from the Soviet government attesting to his leadership in creating the rocket industry and the successful achievement of the Gagarin mission.[5]

Frol R. Kozlov was second recipient of honors. At this time Kozlov was a member of the Communist Party Presidium and Secretary of the Central Committee. He was generally regarded as Khrushchev's heir apparent. A sedate, gray-haired figure, he suffered a stroke in 1963 and died in early 1964. It is important to note, however, that Kozlov received the same honors as Khrushchev. One could thereby infer that Kozlov was Khrushchev's right-hand man in supervising the space effort, acting as the prime link between the party apparatus and the space scientists, dealing with their problems when Khrushchev was occupied with the other difficulties of running the Soviet Union. Kozlov's certificate recognized him for "outstanding services in the development of rocket technology and the achievement of the successful flight of a Soviet man through cosmic space in a satellite ship."[6]

Other high officials were conspicuously honored and written about in the press. These were Konstantin N. Rudnev, deputy prime minister and chairman of the State Committee for the Coordination of Scientific Research (later renamed the State Committee for Science and Technology); Valery D. Kalmykov, chairman of the State Committee for Electronics; and Leonid I. Brezhnev, the current General Secretary of the Communist Party but then the Soviet President. These men received the Order of Lenin for their managerial contributions. The President of the Soviet Academy of Sciences, Mystislav V. Keldysh, and deputy prime minister Dmitry F. Ustinov accepted lesser awards.

Needless to say, the press on that June day was carefully orchestrated. As so often happens, the foreign reporters were excluded and learned of the event through the dispatches of TASS and other media. There was an important reason for the exclusion of foreign correspondents. A number of secret scientists were honored along with the prominent government and party figures. The government newspaper *Izvestia* reported that the military award, the Order of the Red Banner, went to 1,218 unidentified people. The Order of Lenin was given to 478 anonymous workers; the Order of the Red Star went to 256 others; the Order of Merit to 1,789; and various other decorations were scattered around to 3,183 lesser Soviet space technicians and administrators. Yet all remained unknown to the outside world.[7]

Eight years later, in 1969, a brief biography revealed that Sergei Pavlovich Korolyov, the Chief Designer, had been among those present but unmentioned. Soviet President Brezhnev pinned the gold Hammer and Sickle medal on Korolyov's chest and a citation from the Soviet government was read crediting him for "special contributions in the development of rocket technology" and "in the creation and successful launching of the world's first cosmic spaceship, Vostok, with a man aboard." Incidentally, the Soviet journalist who recorded this scene also disclosed that Korolyov had received in 1956 (the year before the long-range tests of the intercontinental ballistic missile and the launching of Sputnik) the title of Hero of the Soviet Union, the Order of Lenin, and a gold Hammer and Sickle medal for his work in military missilery.[8]

The mass award-giving ceremony is notable for the disclosures it made. Yet not much can be learned about the structure and organization of the space apparatus from this sort of Kremlin publicity. The real operating body still lies buried in the unfathomable files, archives, and official secrets, for all the reasons stated by Khrushchev. The student of Soviet affairs, nevertheless, can unearth parts of the organizational structure by combing technical journals and historical documents and by interviewing Western experts who, over the years, have kept a close professional watch on space affairs in the Soviet Union. It has been noted already that in 1954 an Interdepartmental Commission on Interplanetary Communications was established within the Academy of Sciences to coordinate work on an artificial satellite. Several years later, when the Soviet Union joined the International Astronautical Federation, Soviet authorities supplied some more details of the committee's role. It was responsible for developing experiments for projected launches; it maintained contacts with the Soviet population and received applications from Russians who wanted to become cosmonauts; it developed relations with foreign space organizations and sent prominent scientists abroad to attend international space meetings.

A list of twenty-seven committee members was produced, revealing a galaxy of prominent scientists, which included specialists in aerodynamics, gas dynamics, atomic explosives, physics, mathematics, aircraft design, meteorology, astronomy, automation. Even a woman astrophysicist, Dr. Alla G. Massevich, was among them. The first known chairman of the committee was academician Leonid I. Sedov. He was subsequently replaced by a prominent meteorologist, Yevgeny K. Fyodorov, who was succeeded in turn by Lieutenant General Anatoly A. Blagonravov, former president of the Academy of Artillery Sciences. The committee was subsequently renamed The Commission on the Exploration and Use of Space, and has continued to be headed by Blagonravov.

As the years went by, it became clearer that the committee, as important as it might be, was not the primary agency charged with the execution of space flight. In a sense the committee was a front, and Blagonravov was the chief front man. Behind it there existed another agency whose official name is never used, and whose chairman and membership remain secret to this day. The existence of this body has been disclosed in the reportage which accompanied the Gagarin flight and the other manned missions of the Soviet Union. Soviet reporters were allowed to describe the ceremonial meetings of this secret group, known only as the State Commission, before each manned flight at which final formalities were completed.

Foreign observers who have followed the Soviet press accounts will be familiar with the State Commission, or Goskommissiya as it is popularly called. But they will not know much. Its exact mission has never been described, although it appears to bear the final responsibility for all Soviet launches, manned and unmanned. Its lines of communication and cooperation with other Soviet agencies and the military establishment are still unclear, although the Goskommissiya is reported officially to be directly subordinated to the Central Committee and the Council of Ministers. Its true name is hidden, although one source has called it the State Commission for the Organization and Execution of Space Flight. Its membership is uncertain, although some positions have been identified: the chairman, the deputy chairman, the launch director, the Chief Theoretician of Cosmonautics, the Chief Designer, and a swarm of military representatives.[9]

The Chief Designer, Korolyov, surfaces again. This is significant because Korolyov's name was conspicuously absent from those on the roster of the Interdepartmental Commission of the Academy of Sciences. Another fact is still more significant: from a variety of Soviet sources it is known that Korolyov served not only as Chief Designer of the Soviet Union's space hardware, but as technical director of the launch and also as deputy chairman of the Goskommissiya.[10] Government authorities, clearly, had concentrated in Korolyov enormous power and thus it was he who was the leading space genius and possibly, even, the "space czar." According to the continuing number of studies about him, Korolyov was hardly a tyrant. He is pictured as an extraordinary scientist and human being—a man of vision, a hard-headed realist, one who was demanding but sympathetic and charming. To know Korolyov's dreams would be to know a great deal about what Soviet science *hoped* might be accomplished in space.

Korolyov's death must have been a cruel blow to the Soviet program. By the time he died, a sizeable infrastructure of scientists, technicians, and administrators had grown up and were in a position to take over the

machine which had been created. Korolyov had several
deputies who were mentioned from time to time in
newspaper accounts but never identified. Only one,
L. A. Voskresensky, who had worked on rocket development since 1947, has been named—but only after he
died.[11] As far as can be determined, Korolyov's responsibilities were not assumed by any one individual, but
rather were distributed among a group of men who are
said to form a special nucleus within the Academy of
Sciences and the Goskommissiya. Korolyov's proud title
of Chief Designer of Rocket-Cosmic Systems has been
downgraded and refashioned. The Chief Designer today
is merely the Chief Designer of the Soyuz spacecraft
and is involved in the experiments of the Soyuz craft:
orbital rendezvous and the creation of a first generation
orbital station. Korolyov's death probably precipitated a
period of confusion and vying for power within the
Soviet space apparatus, but this cannot yet be documented. It is unlikely to be elucidated for a long time
unless a defector comes to the West with privileged
information which he wishes to disclose.

Despite the great influence and prestige which Korolyov
developed, the formal chief of the Soviet space program
was, and is, the chairman of the Goskommissiya. The
chairman puts his stamp of approval on each mission,
and it is he who bears responsibility on behalf of the
Communist Party for each launch and for the program
in the aggregate. It goes without saying that he is a very
important person in the Kremlin hierarchy, and that his
identity and his exact responsibilities have remained
cloaked in secrecy.

For the first fourteen years after Sputnik, only the
most inconsequential details about the chairman have

been allowed to be published. For example, the Soviet press revealed that the chairman until 1963 was a civilian, an erudite man with a kindly smile. The administrator who replaced him was reported to be a man of less than average height, with a large head planted squarely on a bulging neck. He is said to have a profound knowledge of the problems involved in space exploration and has a gift of explaining them in simple terms.[12]

Soviet censors in 1971 authorized a minor sensation. They permitted the chairman who served during the first manned Soviet flights to uncover his identity in an article, "On the Tenth Anniversary of Man's First Flight in Space." The chairman turned out to be Konstantin N. Rudnev, the same official who had been honored during the earlier, June 20 award ceremony. Rudnev is an administrator with a long history of managerial service in the field of science and military technology. Born in 1911, he completed the Institute of Mechanics in Tula in 1935 and became a chief engineer and factory director during the war years. From 1948 through 1958, he held various managerial posts in defense production, then served, 1958–61, as chairman of the State Committee for Defense Industries. When he became chairman of the Goskommissiya he also was a deputy prime minister and chairman of the State Committee for Coordination of Scientific Research. Rudnev today is chairman of the State Committee for Instrument Building.

On the basis of Rudnev's article (which was circulated to journalists in Washington), it is possible to speculate about two other Soviet leaders who probably play important roles today in the Soviet space program,

Konstantin E. Tsiolkovsky, the venerated Russian
theorist of space flight, in a photograph
taken in 1934, a year before his death. *(Sovfoto)*

The Soviet rocket that launched the first sputniks,
at the launching pad. The picture was released
in 1967, ten years after Sputnik shook the world.
(TASS from Sovfoto)

Friderikh A. Tsander, the Soviet rocket pioneer
of Latvian origin, posed for this formal portrait
in 1913. He was twenty-six. *(The National Air
and Space Museum—The Smithsonian Institution)*

Academician Valentin P. Glushko, an early
rocket pioneer, who is believed to have served
as the Soviet Union's secret Chief Designer
of Rocket Engines. *(TASS from Sovfoto)*

A giant rocket on display during the May 9,
1965, Victory Day parade in Moscow's Red Square.
(*TASS from Sovfoto*)

Yuri A. Gagarin, world's first spaceman, and
Chief Designer of Rocket-Cosmic Systems Sergei
P. Korolyov, whose identity was cloaked in
secrecy for many years, in a relaxed mood.
(Novosti Press Agency)

The augmented Vostok rocket, which has been used to put the latest Soyuz spaceships into orbit. The rocket continues to use the four strap-on stages developed for the original missile that launched the sputniks in 1957 and 1958. (*TASS from Sovfoto*)

Soviet Premier Khrushchev speaking to Valery
F. Bykovsky while the cosmonaut was in orbit,
June 1963. At left is Leonid I. Brezhnev, who
succeeded Khrushchev as head of the Communist
Party in October 1964. *(TASS from Sovfoto)*

but whose connection with the program has not been directly revealed. If Rudnev's own experience is any precedent, it may well be that the current chief of the State Committee for Science and Technology is also simultaneously the chairman of the Goskommissiya. This would be Vladimir A. Kirillin, who served as Vice President of the Academy of Sciences in 1963. He became chairman of the State Committee for Science and Technology as well as Deputy Prime Minister in 1965, and was raised to full membership in the Communist Party Central Committee in 1966.

The other important official in the program is probably Dmitry F. Ustinov, one of the officials who was decorated in the June 20, 1961, ceremony. Ustinov is currently a high Communist Party figure as a candidate member of the Politburo since 1965 and a Secretary of the party's Central Committee. Furthermore, like Rudnev, Ustinov has had impressive experience as an economic, scientific, and military manager. From 1941 to 1945 he was People's Commissar for Armaments, from 1946 to 1953 Minister of Armaments, and from 1953 to 1957 Minister of Defense Industries. In 1963, he became a first deputy prime minister and chairman of the Supreme Economic Council, which was to be abolished after the ouster of Khrushchev in October 1964. Ustinov then began to climb noticeably in the Communist Party hierarchy. It is a fair presumption that Ustinov is now the top Communist Party official responsible for space affairs.

In Rudnev's article there was no exposition of the chairman's personal views on the methods to be used in exploring space. Thus, it is not known whether in the discussion which occurred in the Soviet Union he fav-

ored exploration of space directly by scientist-cosmo-
nauts, or whether he sided with those who emphasized
the advantages and potentialities of unmanned exploration. But perhaps the chairman's influence should not be
overestimated, either. Decisions on such primordial issues as the role of cosmonauts in space and whether to
develop the equipment for sending a manned expedition
to the moon or elsewhere were probably made with the
approval of the highest political levels after careful
consideration of all points of view, including the top
scientists, military men, and financial administrators.

Besides the chairman and the Chief Designer, another
shadowy figure sat on the Goskommissiya. This was the
Chief Theoretician of Cosmonautics. He appeared regularly in the reportage which accompanied the first
manned flights, and was described as on old friend of
the Chief Designer. Journalists reported that he was a
highly competent man, but was reluctant to talk much
about himself. The Chief Theoretician may have been
Mikhail K. Tikhonravov, who, like Korolyov, was active during the 1930s in the GIRD. It was Tikhonravov,
in those years, who designed and flew the Soviet
Union's first liquid-fuel rocket. His name popped up as a
member of the Academy of Artillery Sciences, and he
edited the 1954 edition of the collected works of
Konstantin E. Tsiolkovsky. In recent years, Soviet press
accounts have stopped talking entirely about the Chief
Theoretician, probably because he has left the Goskommissiya and the position has been superseded in the
reorganization following Korolyov's death. Had Tikhonravov died, observers might have expected some
interesting new disclosures.[13]

As to the Goskommissiya's military representatives,

there is little that can be said about them, except the obvious. The Soviet military has played an intimate role in the development of rocketry. Before the Second World War, the highest officers of the military encouraged the rocket pioneers by granting subsidies and taking an interest in their achievements. The war itself gave a boost to the new technology. At first scientists were put to work, as Korolyov was, on developing small jet units to assist the takeoff of aircraft; immediately after the war, efforts began to develop a strategic weapon of a totally new class. Civilian scientists and military resources were joined in a common effort to launch an artificial earth satellite. The Strategic Rocket Troops, the elite service of the Soviet Armed Forces which were created in 1960, have been responsible for the technical work of launching Soviet spacecraft. Western observers have noted that the Soviet military has tested numerous different kinds of space weapons and has developed a variety of intelligence-gathering satellites under the cover of some civilian space programs. Consequently, it is perfectly natural that military representatives should sit on the Goskommissiya and have a significant input in some areas of the over-all space program.

This description of the Goskommissiya fittingly ends with a few observations about secret names. It is not at all unusual in the Soviet press for the identity of authors to be hidden behind pseudonyms. This particularly applies to the many articles which put forth a major concept, explanation, or criticism where the authorities seek purposefully to dilute responsibility by keeping the source vague. The Foreign Ministry from time to time will publish in the press articles signed by "Observer";

Korolyov used occasionally (particularly at the New Year) to publish articles on the space program signed "Professor K. Sergeyev." In the space program there was the extra need to refer to some personalities without actually revealing their true identities. A case in point is the identity of the Chief Designer of Rocket Engines.

A good deal is known about this chief designer's relations with Korolyov. He is described as an old friend from the pioneering 1930s, and a long-time worker in the field of rocket engines. The two men were evidently on close terms, and at one point the chief rocket designer gave Korolyov a small globe of the moon with a note saying he hoped the time would come when they would travel out into space and see a moon of that size with their own eyes.[14] In the history of Soviet rocketry, the person who would seem to fit this description best is Valentin Petrovich Glushko. Glushko worked closely with Korolyov when he was employed at the Leningrad Gas Dynamics Laboratory in developing the ORM series of liquid-fuel rocket engines. Glushko, it is known from early Soviet rocket works, developed the ORM-50 engine, which powered an anti-aircraft rocket designed by Tikhonravov, and the ORM-65 engine, which was destined for a rocket-aircraft designed by Korolyov. In Soviet literature the names of Korolyov, Glushko, and Tikhonravov are frequently linked together, except after the launching of Sputnik when the security veil descended. From then onward, Soviet authors avoided naming the designers when they described the early ORM engines, even though it might have been appropriate to do so.

An interesting phenomenon occurred in the post-Sputnik years, however. By the later 1950s, articles on

rocket engines signed by Professor G. V. Petrovich began appearing. A few references exist which describe
Petrovich as one of the early rocket pioneers and, in
fact, a leader at the Gas Dynamics Laboratory in developing liquid-fuel engines.[15] Checks through the literature of the prewar period talk about Glushko, but never
mention a Professor Petrovich. It seems almost certain
that Glushko turned his name around to form the pseudonym G. V. Petrovich, and that Petrovich or Glushko
was the Chief Designer of Rocket Engines, or, at least,
the chief designer of the early rocket engines such as the
RD-107 and RD-108, which powered the Sputnik
rocket.

On June 26, 1971, another rocket scientist surfaced
from the Kremlin's secret programs. *Izvestia* published
an obituary of Dr. Aleksei M. Isayev, whom it identified
as the director of a scientific team that had developed a
series of rocket engines used on the Vostok, Voskhod,
Soyuz, and other spacecraft. The newspaper did not
specifically label Isayev the Chief Designer of Rocket
Engines, although it implied that he had made very
important contributions.

So well has the secret of Isayev's work been kept that
little of his activities has reached the West. Some minor
historical information on Isayev was collected by U.S.
government analysts and included in the Library of
Congress study *Top Personalities in the Soviet Space
Program* in 1964. Quite typically, this information did
not bear directly on his contemporary role. The *Izvestia*
obituary added a few details. Isayev was born in Leningrad on October 24, 1908, and completed his education
in 1931. He entered the aviation industry in 1934 and
worked on the BI-1 rocket plane designed by V. F.

Bolkhovitinov which flew May 15, 1942. Beginning in 1944, Isayev directed a leading engine development group. The Library of Congress study also showed that Isayev had been one of the original GIRD rocket pioneers but had not occupied a leading role comparable to those of Korolyov, Glushko, or Tikhonravov. On the basis of these facts it seems likely that although he was an important rocket scientist, Isayev was probably not the Chief Designer of Rocket Engines.

An observer could spend an enormous amount of time trying to unravel the mysteries surrounding the Soviet Union's secret personnel and other classified subjects. It is a fascinating exercise but can become diversionary from the general thrust of one's inquiry. Therefore, this account offers only a few more suggestions and these relate to academician Mikhail K. Yangel, academician Vladimir N. Chelomey, and cosmonaut Konstantin P. Feoktistov.

At one time or another, these three names have been suggested by Western observers as possible successors to Korolyov's position as the leading Soviet spaceship designer. Feoktistov's claim is an interesting one. Born on February 7, 1926, he worked with Korolyov in developing Sputnik. In 1964, Feoktistov, whose academic specialty was the physics of bodies traveling through space, became one of the first scientists to fly in a spacecraft when he was launched with cosmonaut Vladimir M. Komarov and Dr. Boris Y. Yegorov aboard Voskhod-1. Subsequently, Feoktistov completed his work for a doctorate in physics and remained active in the space program, as one could judge by the articles and commentaries that he published under his own name. The theory that he had succeeded to Korolyov, however, was proved

wrong by subsequent disclosures. In 1970, United States
space officials who visited Moscow in connection with
U.S.–Soviet talks on the creation of a common spacecraft docking system reported that Feoktistov did indeed
hold a responsible position in the Soviet program. He
was identified as the head of the manned space effort,
rather than Chief Designer.[16]

As to the two academicians, their names have been
mentioned by several Soviet defectors as important designers of space hardware, and possibly contenders for
the title of Chief Designer. One source has contended
that they were rivals of Korolyov's, although at this time
it is impossible to elucidate that claim. As with Korolyov
and Glushko in the late 1950s and early 1960s, little is
known of Chelomey's and Yangel's current connections
with the Soviet space program. But their general biographies have been published. Chelomey, for example,
was born July 30, 1914, and worked, 1941–4, at the
Central Aircraft Engine Institute. He gained a reputation for his work in jet engine designing, and became a
candidate member of the Academy of Sciences in 1958.
It is probably significant that Korolyov, Sedov, Blagonravov, and others known to be connected with space
affairs supported Chelomey in 1962 as a candidate to
become a full academician. He won that honored position in that year.[17]

Yangel is an even more intriguing person. He was
born in a small Siberian village, October 25, 1911, and
graduated from the Moscow Aviation Institute in 1937
and from the Academy of the Aviation Industry in 1950.
He is described in Soviet biographies as a specialist in
an unspecified area of machine building, honored twice
as a Hero of Soviet Labor and once with a Lenin Prize.

He became a member of the Ukrainian Academy of Sciences in 1961 and a full academician in the Soviet Academy of Sciences in 1966. In that year—the year of Korolyov's death—Yangel advanced to an unusually powerful political position. That is the position of candidate member of the policy-making Communist Party Central Committee. Yangel's selection for this high post by the 23rd Communist Party Congress has raised suspicions in the West that he is intimately involved in the direction of the space program.*[18]

These are only tentative guesses formulated on the basis of what defecting and knowledgeable Soviet figures have reported and supported by events in their careers published in small type in the Soviet press. If the past is any indicator, it will take many more years to be sure of the role of these two academicians.

*On October 25, 1971, *Pravda* announced the death of academician Mikhail K. Yangel. An obituary described Yangel as a leading designer in space technology and credited him with important contributions in the Soviet space program.

chapter 4

KOROLYOV

Korolyov came to space flight slowly, but in a logical and gradual way. As a school boy in Odessa in the early 1920s he loved to watch seaplanes taking off from the city harbor. He became acquainted with the pilots, and on occasion even helped them tinker with their craft.

The boy's childhood was a mixed-up one, troubled first by his mother's divorce in 1909 and then, a decade later, by the civil war which wracked Russia and the Ukraine in the years after the Russian Revolution. Korolyov was born in the Ukraine, at Zhitomir, on January 12, 1907. His father, Pavel Y. Korolyov, was a high school teacher whose relations with the boy's mother,

Maria N. Moskalenko, had been shaky for some time. The appearance of a son did not seem to help matters, and the couple separated when Sergei was barely two years old. Mother and child left Zhitomir for Nerzhin, where the boy was consigned to the care of maternal grandparents for some years. Sergei's mother left her parents' house shortly afterward and departed for Kiev, capital of the Ukraine, to study French and to become a school teacher herself. During the next seven years, she visited her son often and, eventually, taught him French. (English and German he picked up by studying on his own.) The year 1916 brought a change in these arrangements with his mother's marriage to an engineer named Grigori M. Balanin. Balanin found a job in that year in Odessa and the family, complete again at last, moved to that Black Sea port city.[1]

As a youth, Korolyov proved to be a good student with a gift for mathematics. Seeking a trade, he entered the First Construction School of Odessa in 1922 and specialized in roofing. He displayed a keen mind and a whimsical sense of humor. Occasionally he would astound students and teachers by hustling down the long corridors on his hands, feet up in the air.

It was during his schoolboy years that he became increasingly interested in aviation, which was developing swiftly in many countries of the world, including Russia. He joined the Odessa gliding club and took a leading role in organizing the society's activities. Gliding proved to be a hobby that would stick with him for many years and that delivered quite a number of professional benefits. It was natural enough, then, that on his graduation in 1924 from the Odessa trade school he tried to get more deeply into aeronautics. He applied to

the great Soviet center of aviation studies, the Zhukovsky Military Air Academy in Moscow, but was judged
to be too young at seventeen and was turned down. So
Korolyov found an alternative route. He joined the
Polytechnical Institute in Kiev and ensconced himself in
the aviation department. Two years later, in 1926, he
transferred to the Aeromechanics Department of the
Bauman Higher Technical School in Moscow, which in
those days was something of an equivalent of the Massachusetts Institute of Technology.

The move to the Soviet capital brought new interests
and stimuli. Korolyov's first introduction to the principles of space flight came in the fall of 1927. While
studying at the Bauman school he noticed a poster ad-
vertising a lecture on space flight by one of the organizers of the famous 1927 space flight exhibit in Moscow.
The exhibit was organized by the Moscow Association
of Inventors and featured models of spacecraft conjured
up by Kibalchich, Tsiolkovsky, and Tsander. There
were also sections devoted to the ideas of Jules Verne,
Dr. Robert H. Goddard, and others. The lecture had its
effect on Korolyov, who then sought out the technical
volumes of Tsiolkovsky and engrossed himself in their
theories and diagrams. Nevertheless, it was still several
years before Korolyov became really deeply involved.
For the time he continued his studies in Moscow, and
actively participated in gliding competitions in the
Crimea.[2]

It is worth noting that Korolyov's interests brought
him in direct contact with a number of Russians who
eventually became leading aircraft designers. Through
his gliding experience he became friends with Oleg K.
Antonov, who fathered some of the Soviet Union's

largest transport planes at the time when Korolyov was designing spacecraft. More significant was his association with Andrei N. Tupolyov, who was already by the late 1920s a noted designer. Through a special arrangement, Tupolyov became the director of his diploma project at the Bauman school. The project was a glider, named the SK-4 (SK for Sergei Korolyov), built with the idea that it might eventually be powered by an auxiliary rocket engine. The craft flew successfully, although in a later flight it crashed, the pilot escaping serious injury. Korolyov suffered great distress from this development, but the professional direction of his life was confirmed. He had been working in an aircraft factory where Tupolyov was employed while studying; now he was certain that he would become an aircraft designer.

After his graduation from the Bauman school the possibilities of rocket propulsion began to interest him more and more. Korolyov was no idle dreamer, but he became affected by the space flight bug. On one occasion he made his own pilgrimage to Kaluga to meet the venerated old man, Tsiolkovsky. Some details of that encounter have been recorded by Korolyov in a little known interview with a TASS correspondent in 1963 which was published four years later in connection with the first decade of space exploration in the Soviet Union:

"One of the sharpest recollections of my life," Korolyov recalled, "was a meeting with Konstantin Eduardovich Tsiolkovsky. I was then 25 years old. We arrived at Kaluga in the morning. We met in the wooden house where the scientists lived. The tall old man met us in a dark suit. During the talk, he pressed a tin ear trumpet to his ear, but asked us to speak softly.

"The talk was not long, but it was detailed. For about
thirty minutes he expounded to us the essence of his
views. I will not vouch for the accuracy of what was
said, but I do remember one sentence. When I, in all of
my youthfulness, honestly announced that henceforth
my goal was to break through to the stars, Tsiolkovsky
smiled, 'This is a very difficult business, young man.
Believe me, an old man. It will require knowledge, de-
votion, and will, and many long years, maybe a whole
lifetime.' "[3]

In his search for more knowledge about the workings
of rocket engines, Korolyov visited many institutions. At
one point he paid a call on the Gas Dynamics Laboratory in Leningrad in 1931. The same year he spent
considerable time in Moscow at the OSOAVIAKHIM,
the Society for the Encouragement of the Aerochemical
Industries. There he met Friderikh A. Tsander, with
whose ideas on space travel he had already become
acquainted. With him, they organized the MosGIRD,
the Moscow Group for the Study of Reactive Propulsion. From the accounts which have come down from
that time it is obvious that Korolyov played an energetic
part. The pioneers sometimes met in Korolyov's own
small apartment in Moscow. The group, whose aspirations ranged from flying higher and faster to actually
leaving the earth for other heavenly bodies, struggled
along at first on its own enthusiasm and resources. In
the first half of 1932, OSOAVIAKHIM recognized
MosGIRD and took it under its wing. And beginning in
August 1932, the Administration of Military Inventions
of the Red Army extended to the pioneers a limited
subsidy.[4]

Within MosGIRD, four sections were formed.
Tsander headed a group which developed primitive

rocket engines. Yuri A. Pobedonostsev worked on a rocket-powered military projectile. Mikhail K. Tikhonravov developed a sounding rocket, and Korolyov sought to adapt rocket propulsion to aviation by continuing his work on rocket-assisted gliders. In November 1933, with Korolyov's encouragement and the backing of Soviet political and military authorities, MosGIRD and the Gas Dynamics Laboratory in Leningrad were fused into a single scientific research institute. Ivan T. Kleimenov was appointed the director of the new laboratory, and on November 9, 1933, Korolyov became deputy director.[5] Characteristically, Korolyov's slogan differed from Tsander's call of "Forward to Mars!" Korolyov declared more practically: "Rockets are weapons and science."

For the next several years, Korolyov busily explored the limits and possibilities of jet propulsion. He constantly consulted Tsiolkovsky's works and corresponded occasionally with the self-taught scientist. Korolyov made two significant appearances, in 1934 and 1935, at scientific conferences that dealt with the scientific uses of rocketry. The first, a meeting on exploration of the stratosphere, was held in Leningrad, March 31 to April 6, 1934, under the auspices of the Academy of Sciences. Here Korolyov spoke about the real possibility of reaching the highest levels of the atmosphere with the help of sounding rockets. His thoughts were developed still further in the address that he gave the following year, in early March, to the First All-Union Conference on the Use of Rocket Devices for Exploring the Stratosphere. Korolyov described in detail the possibility of creating a winged rocket capable of carrying a man aloft for the purpose of taking scientific observations.

At the Jet Scientific Research Institute, Korolyov and his colleagues progressed on two fronts towards the creation of a rocket capable of carrying a man. Korolyov developed a winged rocket, similar to a degree to the German V1, which was launched along a guide rail. It was called the 212 rocket and was powered by the ORM-65 engine developed by Valentin P. Glushko. Of more immediate practical importance, however, were the efforts to create a rocket-powered aircraft. Korolyov joined forces with Ye. S. Shchetinkov. The two designers reworked one of Korolyov's gliders (the SK-9) into what they called the RP-318 (rocket-plane 318). Testing of both the 212 winged rocket and the RP-318 continued with increasing success between 1938 and 1940. Finally, on February 28, 1940, Korolyov's RP-318 was towed into the air and given its first flight tests with a rocket engine. In Soviet aviation history this flight is considered the first of a pure rocket aircraft. It was followed in 1942 by the BI-1 rocket interceptor, developed by another famous aircraft designer, V. F. Bolkhovitinov, later a member of the Interdepartmental Commission on Interplanetary Communications.

Parallel to his work at the Jet Scientific Research Institute, Korolyov was also employed as a designer at an aviation factory beginning in June 1938. He worked directly under his former diploma director, Andrei N. Tupolyov, developing new types of military aircraft and specializing in wing construction. Nearly ten years had passed since he had completed his studies, and he was gaining a solid reputation as a mature engineer. The Second World War forced on him even greater responsibilities. In 1942, he was designated a deputy of Tupolyov's and in 1944 found himself assiduously designing, installing, and flight-testing small rocket units for rocket-assisted takeoff.[6]

The immediate postwar years constituted a crucial period in the history of Soviet rocket development. The Soviet government and Communist Party, as already outlined, began a thorough study of the strategic situation, the Western threat to the Soviet Union, and the possibility of developing a long-range rocket. Korolyov was to play an important part in this development process, but he was to suffer, too. In 1934, he had married a former Odessa classmate of his, Ksana Vintsentina, and by the war's end it had become obvious that it was an unhappy union. More threatening still, in 1945, Korolyov was taken, on unspecified charges, and confined to a special work camp. The details of his imprisonment are admittedly sketchy and have never specifically been mentioned by Soviet authorities, no doubt because of their embarrassing connotations. It was a fate that befell thousands of loyal Soviet citizens during the xenophobic late years of the Stalin regime. Korolyov's confinement was disclosed by one Soviet defector in a position to have at least some knowledge of the designer's background. Furthermore, one Soviet biography of the future Chief Designer indicates that he worked in isolation on rocket boosters immediately after the war, and even quotes from a letter to his mother describing his ascetic living conditions.[7] Apparently, Korolyov's performance, in this trying and despairing situation, was noticed. Toward the end of 1946, Korolyov was named to head one group of scientists working on rocket development. He took up his duties in February 1947.[8]

Most of the knowledge available in the West about the Soviet government's decisions regarding long-range rocketry in 1947 has been passed on by Tokaty-Tokaev. Only recently—twenty-five years afterward, to be exact —has a brief glimpse of a meeting between Stalin and Korolyov surfaced. Korolyov's description of this encounter does not reveal a great deal, but it does tally with Tokaty-Tokaev's recollections of Stalin's intense interest in creating a transoceanic rocket.

"I had been given the assignment to report to Stalin about the development of the new rocket," Korolyov remembered. "He listened silently at first, hardly taking his pipe out of his mouth. As he became interested, he began frequently interrupting me, asking short questions. One got the impression that he had a full understanding of rockets. He was interested in their speed, their range, and altitude of flight, the payload which they could carry. He kept asking, with particular interest, about the accuracy of the rocket's flight to the target. Stalin, on the surface, was restrained. I did not know whether he would approve what I was talking about or not but this meeting played its positive role. Apparently Stalin and his military advisers understood that the first attempts at building jet aircraft and rocket systems could produce positive far-reaching results, later on."[9]

There are still many details to be filled in about the development of long-range rockets in the Soviet Union in the decade from 1947 to 1957. Some high points, of course, are known and have been briefly mentioned. Groups of German scientists were imported and put to work in isolation from Soviet specialists under the general supervision of Dr. Yuri A. Pobedonostsev. The Gas Dynamics Laboratory in Leningrad continued its experiments with liquid-fuel rocket engines, and between 1954 and 1957 successfully developed the RD-107 and RD-108 engines which are still being used today. As for Korolyov, Soviet sources acknowledge that his full energies focused on a truly long-range rocket beginning in 1953, the year in which he was raised to new scientific standing by his election to the position of corresponding member of the Academy of Sciences.[10] This disclosure also accords with speculation, derived from external evidence, that during the last year of Stalin's life, 1952–3, the Soviet government had reached a critical moment of decision in its rocket development plans.

During the years when the rocket vehicle was under development there was practically no public evidence that the secret scientists existed or that thought was being given to the possibility of space flight. But it can now be said with certainty that the implications of the new technology for scientific purposes were being actively discussed at various levels. And this was not so unusual. Tokaty-Tokaev recalls that before the war he had received Korolyov in his office at the Zhukovsky Air Academy and that the future Chief Designer had declared that Tsiolkovsky's notions on space travel were becoming realities.[11] This was a theme which Korolyov would repeat to those who would listen—and particularly in 1947 on Tsiolkovsky's ninetieth birthday. On that occasion, September 17, 1947, a memorial meeting was held in Red Army House under the auspices of the Academy of Artillery Sciences. *Pravda* reported in a few brief paragraphs the following day that Lieutenant General Anatoly A. Blagonravov, the President of the Academy, opened the gathering and discussed Tsiolkovsky's theoretical contributions to artillery. The *Pravda* article also noted that an S. P. Korolyov, identified only as a corresponding member of the Academy, delivered a report on Tsiolkovsky. To even the most assiduous Western observer at the time, the notation meant little. Korolyov's report was never published. But, in 1969, some excerpts did become available that show Korolyov's continuing preoccupation with scientific adaptations of the new technology:

"He [Tsiolkovsky]," Korolyov said, "came to analyze in detail the question of the flight of rockets. He produced a series of brilliant technical solutions for rocket flight which carry to this day his name and are used everywhere. Among these is the relation of the speed of movement of the rocket and the speed of the discharge of the combustion products and the logarithm of the relation of the end-mass and the initial mass of the rocket [Tsiolkovsky's Formula]."

More interesting still:

"He developed the project of the creation of an artificial satellite of the earth in the nature of a temporary stage or station, which should be created on the road to cosmic voyages. This miracle is a fantastic and impressively grandiose idea even in our day, but one must recognize that this is a scientific truth and a scientific forecast not so far removed in the future."[12]

A decade later, on a similar occasion, Korolyov was again principal speaker. This time Tsiolkovsky's 100th anniversary was being celebrated not in military surroundings, but in the scientific headquarters of the Academy of Sciences. Korolyov's speech was only briefly summarized by the Soviet press at the time. An abridged version of his remarks has since been made available by the Soviet National Association of Historians of Natural Sciences and Technology. (How Western intelligence analysts would have appreciated a copy at the time!) Contrast the changes in Korolyov's language; the details are meager, but the changes are significant:

"The Soviet Union is successfully testing super long-range intercontinental, multistage, ballistic rockets. The results obtained show that it is possible to launch rockets to any region on the earth. During the period of the now approaching International Geophysical Year, dozens of rockets will be launched to conduct scientific investigations according to a diverse program for different altitudes and in different regions of the Soviet Union, including regions of the far north and at Soviet expeditions to the Antarctic.

"In the near future the first launchings of artificial earth satellites will be carried out in the U.S.S.R. and the U.S.A. Soviet scientists are working on many new problems of rocket engineering; for example: on the problem of sending rockets to the moon and a lunar fly-by; on the problem of manned flight in a rocket; on the problems of deep penetration into and investigation of space. The remarkable predictions of Konstantin Eduardovich Tsiolkovsky concerning rocket flight and flight into interplanetary space which he expressed more than 60 years ago are coming true."[13]

Korolyov's extraordinary statement reportedly elicited no wild surprise. The audience received the news in silence. Yet in historical perspective his remarks can only have meant that Soviet scientists were already seriously engaged in planning and preparing the startling achievement that they unfolded so competently between 1957 and 1961: the first artificial satellites in 1957; the first rocket impact on the moon and photographs of the

moon's hidden side in 1959; man's first orbital flight
in 1961. By the mid-1950s, Soviet scientists were developing a multistage rocket with a projected thrust of 1.1
million pounds and they were examining the many uses
to which this device could be put—both military and
scientific. This did not mean that the plans for exploring
space were universally approved; quite the contrary.
There were scientists in Russia who mistrusted the
efforts; there were others who believed in the possibility
of accomplishing the space aims but who doubted that
the expenditure was worthwhile. Possibly because of
this division of scientific opinion, reported by Soviet
sources today, the highest political authorities maintained a reserved attitude. Nikita S. Khrushchev, the
First Secretary of the Communist Party, did not attend
the launching of the historic Sputnik; neither did he
immediately engage in effusive public exclamations
when the news was relayed to him. The reserved attitude of the Soviet leadership at the beginning of the
space venture possibly accounts for the fact that the
announcement of the Interdepartmental Commission on
Interplanetary Communications was so low-key, and,
for Western observers, so very easy to dismiss.[14]

Just as U.S. experts tried to follow Soviet rocket
developments, so the Russian specialists watched the
United States. The plentiful reporting of U.S. efforts to
orbit a satellite in connection with the International
Geophysical Year was readily available to Korolyov and
to others. It was easy for them to calculate how much of
a satellite the United States could put into orbit if the
U.S. Navy's Viking rocket—the designated American
booster—worked successfully. And Korolyov's group
did just that sort of calculation. Korolyov reported to

the Central Committee apparatus that the Soviet Union disposed of considerably more rocket power in its proto-type intercontinental missile than the United States did in its Viking rocket. Korolyov added that there was every reason to believe that the Soviet Union could orbit a satellite before the United States, and what is more, could send up a significantly greater payload. The Central Committee response to this intelligence, according to one source, was: "That is a tempting business, but we'll have to think about it."[15]

It cannot yet be pinned down exactly when Korolyov made his proposal to beat the United States into earth orbit, but it is likely that this occurred in 1956 or before. What is known is that during the summer of 1957, presumably following the successful rocket tests that extended from July through August, Korolyov was summoned to the Central Committee and told to go ahead. The designer and his group then departed for the Baikonur Cosmodrome—the Soviet version of Cape Kennedy—to begin the final work on preparing "the simplest sputnik" for launch.

One can easily believe that the success of Sputnik had its effect on both scientists and political figures in Russia. For Khrushchev there was much to shout about. The launch came only weeks before the Soviet Union celebrated the fortieth anniversary of the Bolshevik Revolution, and the First Secretary in press interviews and speeches could draw plenty of attention to the forward march of socialism. As to the scientists, on Korolyov's insistence they rejuggled their calculations. The plans developed in the mid-1950s called for experiments laying the groundwork for a flight of man into earth orbit. The orbiting of an experimental animal had

been planned as the third launch, and the necessary
cabin and life-support system had been readied accordingly. Korolyov now insisted on advancing the launch
of a dog, and so, only a month after Sputnik, the Soviet
Union on November 3 orbited Sputnik-2 with Laika on
board. Sputnik-2 weighed 1,120.1 pounds and proved
to Western observers that the Soviet Union had developed a remarkably powerful rocket and seemed to be
absolutely serious about putting a man into space, thus
going far beyond the scientific goals of the I.G.Y. There
is a lingering question about Sputnik-2: Did Korolyov
scrap earlier scientific plans and push forward Laika's
flight on his own? Or did he do it for political reasons,
on the insistence of Nikita S. Khrushchev?

The successful start of a space research plan seems
to have resulted in greater backing for Korolyov's group
from the Soviet government and the Communist Party.
This cannot be ascertained exactly, but it appears that
following Sputnik a more formal and permanent apparatus was approved for managing the exploration of
space. It was, of course, the State Commission for the
Organization and Execution of Space Flight, otherwise
known simply as the Goskomissiya. The commission, as
already noted, was made responsible directly to the
Communist Party Central Committee and to the Council
of Ministers of the Soviet government. The group was
built largely around Korolyov. He held simultaneously
the positions of deputy chairman of the commission,
Chief Designer of Rocket-Cosmic Systems, and technical director of flights. He and his technical council, in
consultation with appropriate scientists, drew up long-
range plans and developed the hardware to perform
the tasks. The commission then passed on the plans in a

formal manner, and took the responsibility for executing them. But in the early years the responsibility was Korolyov's to a very large extent.[16]

Thus it was that the project of launching a rocket to the moon originated first with Korolyov and his closest colleagues. His group carried out various studies of flight paths to the moon, around its hidden side and back to earth, at least as long ago as 1957 and quite possibly before. When the Korolyov group had reached the conclusion that the flight was feasible, an expanded meeting of scientists was held to discuss how the hidden side of the moon could be photographed and relayed back to earth. The expanded group included such well-known Soviet astronomers as Alla G. Massevich (the woman member of the Interdepartmental Committee), N. P. Barabashov, V. V. Sharonov, and A. A. Mikhailov. They were reportedly astounded that Korolyov envisaged a scientific probe of the earth's natural satellite so soon. Characteristically, the first lunar probe was named "Mechta"—The Dream—and was launched January 2, 1959, by the same rocket which had boosted the first sputniks into orbit (augmented with a small upper stage). The probe subsequently became labeled Luna-1 and formed the beginning of a lengthy series that has extended up through 1970 and has included a large number of assignments such as fly-bys and soft landings on the lunar surface. But the term "The Dream" was clearly a reference to Tsiolkovsky's hopes for lunar exploration, if not to man's general fascination through the ages with the moon.

The development of lunar probes early in the Soviet space program was a special dream for Nikita Khrushchev. On January 4, 1959, "The Dream" sailed by the moon at a distance of six thousand kilometers (3,720 miles) and veered into solar orbit. Nine months later, Korolyov and his colleagues launched Luna-2. On September 12, it slammed into the moon's surface, scattering a payload of emblems bearing Soviet insignia. And then, most sensational of all, on the second anniversary of Sputnik a third lunar probe circled the moon, photographed the hidden side, and transmitted those pictures back to earth.

These were glorious moments for Khrushchev. He never tired of boasting in those years of Soviet prowess in shooting the moon, and of the continuing failures that the United States was suffering with its Pioneer series of lunar probes. But Khrushchev's very special joy came in mid-September 1959, on his only visit to the United States. One of the first things he was able to do upon arrival was to present President Dwight D. Eisenhower with replicas of the medallions that Luna-2 had deposited on the moon. It was a gesture full of the symbolism of power.

The year 1959 saw some very impressive Russian achievements, but the lunar exploration program lapsed until 1963, when moon shots were resumed. This was odd because the Soviet scientists had convincingly demonstrated their abilities, and each month they were presented with an optimum period for aiming at the moon. One is tempted to the conclusion that scientific endeavor was not the entire motivation of the early Luna flights. Had Khrushchev participated in the decision-making of the moon flights, exerting his influence for a few immediate, sensational technological achievements rather than for well-prepared but more distant scientific missions?

Korolyov surely was proud and encouraged by his success in opening up the new space era with the help of a single military-scientific rocket. On the other hand, he was hardly a frivolous personality, and is portrayed as rejecting the notion that the Soviet Union was merely seeking to establish records in space and collect a list of spectacular "firsts," as has been suggested abroad. Nevertheless, he struck a balance between caution and bold, attainable steps. For example, he opposed those Soviet scientists (still unnamed) who are reported to have questioned the wisdom of sending the first cosmonaut into a full orbit of the earth. A group of scientists in the Soviet Union favored the more conservative approach of experimenting, first, with suborbital flights for human beings, as was done in the United States. Korolyov buttressed his position with a scientific paper that sought to define cosmic flight as requiring orbital velocity (or greater) at an altitude well above the dense layers of the atmosphere, and for a "fairly lengthy period" roughly equivalent to one revolution around the earth. This guideline was accepted.[17]

Recruiting for cosmonauts began in 1959. Yuri A. Gagarin, who was to be chosen for the first flight, reported that he was inspired by the first moon probes and applied forthwith to become a cosmonaut. Presumably, by then, Korolyov's design group was well along in the creation of the spherical Vostok spaceship. The year 1960 was largely given over to flight tests of this craft and its landing apparatus. Precursor flight tests were held on May 15, 1960, August 19, 1960, December 1, 1960, March 9, 1961, and March 25, 1961.

It is interesting to note that the first of these flights was a technological failure but a political success. On May

1, 1960, Soviet rocket defense forces near Sverdlovsk managed to down the U-2 spy plane, piloted by Francis Gary Powers, which was overflying the Soviet Union. The incident precipitated a deep crisis between Khrushchev and Eisenhower, who, at that time, were completing preparations for a summit conference in Paris with British Prime Minister Harold Macmillan. The East–West summit—first since the 1955 Geneva conference —eventually broke down over Eisenhower's refusal to apologize directly to Khrushchev for sending the spy plane over his country. Meanwhile, the May 15, 1960, Vostok precursor flight brought some welcome news for the injured party. The world press began to speculate that the flight was directly preparing a manned mission in the fairly near future. The Vostok precursor carried some minor passengers—mice and flies—but it was intended to return to earth. When the scientists turned on the braking engine, a malfunction caused the ship to zoom into a higher orbit that kept it revolving around the earth for 844 days. The following four precursors, carrying either dogs or dummies or both, were successful and gave the scientists every confidence for Gagarin's April 12, 1961, historic solo.

The details of Gagarin's one-orbit flight, and the one-day flight of his backup pilot Gherman S. Titov on August 6, 1961, have been recorded in considerable detail. Carefully edited news dispatches were released at the time. Upon their return both cosmonauts wrote their impressions in popular books. Soviet scientists have revealed many of their findings although they have withheld others. It was only recently, however, that details became known of an argument between Korolyov and other scientists in their planning of the Titov flight. Lieutenant General Nikolai P. Kamanin, the cosmonauts' trainer, and one of the leading space doctors, V. I. Yazdovsky, urged that Titov be launched on a three-orbit mission (similar to what would be the flight of U.S. astronaut John Glenn in 1962). They, and others, apparently did not want to commit Titov to a flight of some duration, particularly since the opportunity to land on Soviet territory existed only on the early orbits. But Korolyov objected. He insisted that Titov be sent off for a full day in space, arguing that it was esssential that a cosmonaut fly for a prolonged period and attempt to work in space. The details on this disagreement are still most fragmentary and when the full story is told, if it ever is, there may be cause for some surprise.[18] Did Korolyov really overrule the medical specialists entirely through his own confidence and conviction, as Soviet sources are currently suggesting? Or did Khrushchev apply pressure for a longer, more impressive flight?

It is unfortunate that so little is known from authoritative sources of Khrushchev's relations with Korolyov. There are enough rumors about, some of them in print, which would suggest that Khrushchev was demanding of the scientists' efforts. Publicly, Khrushchev used to suggest that he, like any other political figure in any other country, was really at the mercy of the scientists. And yet, at the time, *Pravda* and other sources encouraged the notion that he participated in "the determination of the basic directions and establishment of generally planned growth of cosmic science and technology." In his book *Road to the Cosmos,* Gagarin quoted Korolyov as saying that space exploration was Khrushchev's pet subject. "He [Korolyov] spoke of his meetings with Khrushchev at the Central Committee," Gagarin wrote, "in laboratories and at the Cosmodrome. He added that Khrushchev devoted a great deal of care and attention to this new sphere of activity."[19]

One flight in which Khrushchev may have wielded a direct influence was the group mission, June 16–19, 1963, of cosmonaut Valery F. Bykovsky and the world's first woman cosmonaut, Valentina V. Tereshkova. Some months earlier, rumors circulated in journalistic circles in Moscow that a woman was soon to fly and that the purpose of the flight was chiefly propagandistic. Korolyov has commented (as will be shown shortly) on the launching of a woman cosmonaut. He stressed that it would prove the courage of Soviet women and their equality with Soviet men. The launching of a woman without pilot experience would also prove that the Russians had developed a reliable method for preparing cosmonaut candidates, and that they possessed reliable, automated spacecraft. Such a mission would have significance for later flights, when scientists without extensive flying experience would be sent into orbit.

In historical perspective, several facts stand out about Tereshkova's performance. Since Soviet cosmonauts began flying in 1961, Tereshkova has been the only woman to fly. None have followed her, and none at this point seem destined to, even though a group of women cosmonaut-candidates are reported to be in training. Secondly, after Tereshkova's return she was quickly presented to the International Congress of Women, which was meeting conveniently in the Kremlin. She instantly became a Soviet-style glamour girl. There remains a fairly strong suspicion that Tereshkova's flight—like other aspects of the Soviet space program—

had a double purpose: scientific and pure propaganda.

But what about the most sensational of all possibilities in the space program—the flight of Soviet cosmonauts to the moon? In the late 1950s and early 1960s, Soviet scientists offered numerous comments on the project, which added up to a determination to launch such a mission as soon as it was technologically feasible. One of the most authoritative to speak on the matter would be Chief Designer Korolyov, of course, and in 1963 he did so in a particularly interesting interview. At the time an edited Russian version of his comments was published by *Izvestia* while only fragmentary bits were circulated abroad in English by TASS. Furthermore, the remarks were attributed only to the Chief Designer, and released at a time when Korolyov's identity was not known and when the nature of his position of Chief Designer was only vaguely understood abroad. The interview was made during the group flight of Bykovsky and Tereshkova and transmitted in a highly edited version June 22, 1963; a fuller version has now become available:

Question: What new elements have been brought to cosmonautics by the launching close together of the ships Vostok-5 and Vostok-6 in comparison with the group flight of Andrian Nikolayev and Pavel Popovich [in August 1962]?

Korolyov: This group flight of spaceships has been calculated for a lengthy period. The new probe into the cosmos has been motivated by the necessity of gathering experience for lengthier orbital flights, and for the creation in orbit around the earth of automatic stations, and, in the long run, for reaching the planets. It's worth recalling, for example, that the flight of the interplanetary automatic station Mars-1 [launched November 1, 1962] took seven months, and that the American probe Mariner 2 took about four months to the closer planet of Venus. In fact, a flight to the closest natural satellite of the earth, the moon, or more correctly, a flight around it would take six to eight days.

The second group flight, like the first, is engaged in acquiring experience. It's worthwhile to test all equipment in conditions of flight, master all the many operations of the ship, and systems, which insure the lives of the crew. These problems are extraordinarily important, as is the growth of experience in piloting ships, their orientation in space.

Finally, there is another very important problem of a biological character. We must know how man will support weightlessness, particularly if he is to be in this condition for a long period. Acquired experience allows one to assert that in the period of the first days of a flight, the organism begins to adjust to weightlessness.

In this connection the flight of a woman into the cosmos, Valentina Tereshkova, and her participation in the group flight with Valery Bykovsky is of special interest. Until recent times, cosmonauts were exclusively jet pilots, trained personnel, accustomed to gravity overloads and to speed. The commander of Vostok-6 ran into g-loads only in the process of training for a cosmic flight.

If we can say so, without the exaggerated claims which we so often use, we could evaluate the first flights of man in near earth space this way: Gagarin's flight was a first serious probe. Titov's flight was a deep probe. The flight of Bykovsky and Tereshkova is still another step forward, both in the sense of lengthiness and in the scientific, technical questions before us.

Question: What other scientific goals, besides those mentioned, were there for this flight?

Korolyov: There were a series of astronomical observations of the constellations. Many photographs of the sun are planned. At sunset and sunrise he [Bykovsky] is to photograph the changing spectrums. The value of these photographs is that they will be taken without interference by the atmosphere. The measuring of the radiation belts of ionized particles at the altitudes at which Vostok-5 and Vostok-6 are flying has particular significance. Observations of the earth—both visual and with the aid of optical instruments—are very important.

Question: Do the three-day and four-day flights of the cosmonauts prove that the fears of the skeptics are groundless: weightlessness is not so worrying as it had been depicted?

Korolyov: I am a convinced optimist and believe that very prolonged interplanetary flights of man are not so far off. But nevertheless weightlessness, and its effects on the organism of man, are far from being completely understood. There, if you please, simple optimism is not enough. Large collectives, headed by leading scientists in biology and medicine, are studying these problems. We expect decisive answers from them.

I want to develop the thought why a cosmonaut has to be able to "read" the geography of the earth. It is indispensable for a cosmonaut to know his home planet by the characteristics of the earth on return from lengthy flights. It follows that one must learn accurately to distinguish what the mountain ranges and snowy peaks look like from space. You will ask: how about navigational instruments? That's all very well. Nevertheless, the eye of a cosmonaut, possibly aided by special optics, is a very fine instrument. And remember: it's better to see once, than to hear a hundred times.

Question: You've said already that a flight, where the commander is a woman, is an important event. What else did you want to add to that?

Korolyov: It is first of all one of the sharpest proofs of the equality of Soviet women and their great courage. I should say that women sent many letters with the request that they be accepted into the group of cosmonauts. Secondly, the flight of women in space is proof of the high level of our technology. This technology is manageable not only by an experienced pilot, but also by a person who does not have flight experience. Now it is clear that the problem of training cosmonauts is solved. But, of course, one should not imagine that to become a cosmonaut is easy. In order for a cosmonaut successfully to fulfill his tasks, it is necessary to undergo serious and prolonged training.

Question: The spaceships Vostok-5 and Vostok-6 flew in close orbits. What significance was there in their proximity?

Korolyov: That is a big and very important problem—the problem of rendezvous and joining, or as we say, docking of spaceships. It is on the program of space flight. Its solution will produce a lot: it will be possible to build large orbital stations which would be able to serve research goals and at the same time constitute their own special kind of docks for spaceships.

I would compare space flight to seafaring. Spaceships, like ocean-going vessels, leave land for a long time. In order to stock up on necessary supplies, either one ship or the other could return to earth or seek a rendezvous along the way. Accordingly, group flights bring us closer to the solution of this problem. Up until now spaceships have been flying on so-called self-braking orbits. That means that in any case, even when the braking system fails, the flight of the ships will be slowed by the atmosphere, and they will descend to earth in a relatively short time. The presence of space docks, the possibility of joining ships will permit the use of higher orbits, and will boost the boundaries of space travel.

Question: Will the joining or docking of objects in space require man to leave the confines of his ship?

Korolyov: Unquestionably, but not like the man who links up carriages at railroad stations. I am a supporter of the orbiting of parts, for example, of space stations into the cosmos and the automatic joining of them there in a unified complex. Man would fulfill the duties of a dispatcher, a supplementary controller, and, maybe, somewhere would intervene in the work of the automatic devices if they should operate inaccurately.

Question: Speaking of the widening of the boundaries of the flight of spaceships, do you have in mind the reaching of the area of the moon by man?

Korolyov: Undoubtedly. But one should say that the flight, for example, to the moon is extraordinarily tempting but a very difficult problem. A flight to the moon is complicated, the return to earth even more so.

Soviet scientists are working on the solution of the problems involved here. I'm sure that the time is not far off when the journey of man to the moon will become a reality, although more than one year will be necessary for the practical solution of this problem.[20]

There are several interesting elements which emerged from this 1963 interview. The first was the definite conviction that man would fly to the moon, or, at least, circumnavigate it. Korolyov said this would happen at a time which was not far off. To the Anglo-Saxon ear, this sounds as if the event were reasonably near; say, in the next couple of years. Yet the Russians use the expression—*"ne daleko to vremya"*—in a much more indefinite sense than an English-speaking person would. The Chief Designer followed up the indefinite timing of the moon project by stating further that "more than one year will be necessary for the practical solution of this problem." This is a cultural difference worth noting, because many Western observers have tended to interpret Russian statements on a manned moon flight "not being far off" as meaning they could occur shortly.

The Korolyov interview referred to the creation of a large earth orbital station which would serve as both an observation post and as a staging area for manned probes into the solar system. As the years have gone by, Soviet authorities have talked more and more about the creation of orbital stations, and less and less about direct flights of men to the moon. It is significant that, in 1963, Korolyov seemed convinced an orbital station should be created to form a necessary stopping-off place on the way to the moon. Such a way station would have certain advantages. If a lunar expedition were assembled on an orbital station outside the earth's atmosphere and dispatched from that point to the moon, the lunar spaceship would require a far less powerful engine than if it were launched from the earth, as Apollo 11 was. Furthermore, the orbital station would have uses as an observation point from which to survey the earth, as well as the rest of the solar system. But creating such an orbital system would take time. In June 1971, the Russians created a first-generation orbiting station by linking up a Soyuz ship with a Salyut module. Surely Korolyov must have realized the long lead times and delays that would be involved in perfecting docking techniques, fueling procedures, and creating an exotic engine for the lunar craft. He must have suspected, therefore, that a flight mode using this method would not be competitive with U.S. plans for landing men on the moon by the end of the 1960s.

Soviet authorities today are creating the impression, through the release of previously unavailable documents, that the current Russian emphasis on creating manned orbital stations is the direct outgrowth of Korolyov's thinking in the early 1960s. These officials may be withholding other evidence which would show that Khrushchev—or another group of influential but secret scientists—urged the creation of a large rocket vehicle for the purposes of making a direct ascent to the moon. U.S. intelligence, starting in 1964, began receiving evidence that the Russians were building such a giant booster. This rocket, whose existence has been revealed by James E. Webb, former director of NASA, on the basis of American intelligence, is said to develop 10 million pounds of thrust in comparison to the 7.5 million pounds of thrust of the American Saturn V. What role may have been projected for this huge space vehicle—whose existence has never been acknowledged by the Russians—will be discussed in a later chapter.

Some further inkling of Korolyov's thinking during the last years of his life has been identified in articles that he wrote at the beginning of each new year, starting January 1, 1964. These articles, published by *Pravda*, surveyed the achievements of the last year and looked into the future. At the time it was impossible to assess the authority of these articles because censorship prevented Korolyov from signing them. Nor were they designated as having been composed by the Chief Designer. Rather, they were signed simply "Professor K. Sergeyev."

Years after his death in 1966, it became known that Professor K. Sergeyev was really a pseudonym for Sergei Pavlovich Korolyov. There are various references to this subterfuge—one, for example, in a posthumous edition of Yuri Gagarin's *Road to the Cosmos*.[21] All references pass off the pseudonym in a casual manner, as if trying to call as little notice to it as possible. None of the references so far have called particular attention to Korolyov's practice of forecasting in *Pravda* what the coming year would bring in space flight.

Here are some of the high points from these articles. In the January 1, 1964, edition of *Pravda* Professor Sergeyev predicted that cosmonauts would continue orbiting the earth, dampening any speculation that a moon flight might be accomplished in the foreseeable future, or even by the Soviet Union's fiftieth anniversary in 1967. Yet he wanted to keep open the possibility that flights to heavenly bodies would be mounted when possible. He reasserted, as he had in 1963, that undoubtedly cosmonauts would fly to the near planets of Mars and Venus as well as to the moon. The time was not far off, he said, when spaceships returning from long flights would tie up at "cosmic docks" and the cosmonauts would rest in comfortable orbiting stations.

In the edition of *Pravda* published January 1, 1965, Professor Sergeyev said that it would be necessary for man to learn how to perform extravehicular activity in space. He had made this point in the 1963 interview, but the significance of the 1965 mention was that it preceded by several months the actual event. It was a hint so guarded that Moscow correspondents and other close observers failed to pick it up. And then on March 18, 1965, Voskhod-2 was launched and, via video-tapes and television, the world got its first view of cosmonaut Alexei A. Leonov emerging from the spaceship, tumbling and somersaulting in the cosmic void.

Professor Sergeyev touched on other matters, too. Again he emphasized the value of comfortable, habitable stations circling the earth where scientists of many different specialties could work. He also introduced a new idea which may have had some important but veiled significance at the time. Technical solutions to problems of space exploration, he said, must be made with an eye to reliability and economic considerations. The emphasis on the costliness of space exploration was a new departure. Any informed observer knew that space exploration was costly, and that manned exploration was particularly so. Soviet personalities occasionally remarked on this, but on the whole the Soviet government tended to minimize emphasis on the expense according to statements which appeared in the press. The Soviet space budget still has never been officially disclosed.

Finally, in *Pravda* of January 1, 1966, Professor Sergeyev came straight to the point of a moon-landing expedition. "There is no need to say," Korolyov wrote, "how long, how strongly, how relentlessly the moon has caught the attention of man. The dream of humanity has been the desire for a child of the earth at last to land on the lunar surface. Unfortunately, this task is not such a simple one, and not so close to achievement.

"The moon is the natural and eternal satellite of ours. It has substantial differences from the earth. There is no atmosphere on the moon. There also are no appreciable magnetic field or radiation belts.

"In these unusual conditions, existing only on the moon, there are great possibilities for scientific research which are absolutely unattainable on earth."

This was Korolyov's last article. He finished it in December 1965 and read it to Nina I. Korolyova, his second wife. Then he left for a Moscow hospital and the fatal operation.

The Chief Designer had had periodic difficulties with his health as far back as Sputnik days and possibly before. Biographic information available on him presently shows that he had bouts of heart trouble and intestinal ailment in 1957, 1961, 1962, and 1964. In June and July 1964, he and his wife made their only known trip abroad—to Czechoslovakia. Korolyov was supposed to take a cure at the Czech spa of Karlovy Vary, but, in typical fashion, he refused to stay still. The couple traveled around the country visiting places of interest to him.

Korolyov underwent medical tests in a Moscow hospital, December 14–17, 1965. After the New Year— one of the most important Soviet holidays—he was admitted for the operation on January 14. At one point he asked his doctor how long he could expect to live with his battered old heart.

"Well, I think another twenty years," the doctor replied hesitatingly.

"I would be satisfied with ten, although there is still a great deal to do," Korolyov replied.

The operation proved difficult. The surgeons struggled over Korolyov for hours. When Nina arrived to see him after the operation, she was greeted with a shock: "Take courage," said one of the surgeons, "it is all over."[22]

In its issue of January 16, *Pravda* reported Korolyov's death two days before his 59th birthday. For the first time he was identified as the Chief Designer of Rocket Cosmic Systems. An accompanying medical report stated that he had been suffering from a malignancy in the main intestine, sclerosis of the arteries, emphysema, and an upset metabolism. The immediate cause of death was said to have been "cardiac insufficiency" during the operation.[23]

chapter 5

KHRUSHCHEV AND THE SPACE RACE

An Italian industrial fair was opening in Sokolniki Park. This wooded area in northeastern Moscow was frequently used by the Soviet authorities as a site for foreign exhibitions. And Sokolniki had a growing reputation; besides being a vast tract of forested land divided by seemingly endless, winding paths, it was a place where Russians could get a glimpse of fascinating foreign products, talk with amiable guides, and rub shoulders, from time to time, with visiting statesmen. It was here that the American exhibit of 1959 was held, and where the then Vice President, Richard Nixon, and Khrushchev got into a verbal slugging match that became known as the "kitchen debate" because it occurred in the model

American kitchen. Now, on May 28, 1962, foreign correspondents were again trooping down the gravel alleys,
hoping that the ebullient, unpredictable Khrushchev
would turn up.

To reporters based in Moscow, Khrushchev was a
journalistic gold mine, and every appearance he made
had to be covered. His journalistic value was such that
correspondents spent a fair amount of time trying to
figure out when and where he might make an impromptu
appearance. For Khrushchev was the one Soviet official
who had the authority and inclination to be indiscreet,
informative, and frank. Beyond that, he was responsive.
He had come to recognize a number of the Western
correspondents in Moscow, and could be generally
relied upon to answer questions of burning interest. In
short, Khrushchev was a splendid news source, if and
when you could get to him. In the history of the Soviet
Union he was probably the best public relations expert
since the heady days following the 1917 Revolution,
when Lenin was reasonably accessible to Westerners
and when there existed a measure of openness in Soviet
politics that has long since disappeared.

Aside from these qualities, Khrushchev was colorful
and intensely human. Unlike the traditional Soviet official, who avoids talking in public about his private life,
Khrushchev gloried in recounting his past and empha-
sizing the superiority of a social system which had
allowed him to become one of the world's most
powerful men. He would recall, on occasion, how he
had been born in the little village of Kalinovka near the
Ukrainian border in 1894 during the era of the big bad
capitalists in Russia; how he was allowed only two
years of formal education; and how he and his father

struggled to make ends meet by alternately working the mines and the fields. He recalled his work as a metal fitter at Yuzovka, a town in the Donbass named after a Welshman, John Hughes, who started the mining and metal industry. And he reflected with a certain pride how he progressed from a youth shod in birchbark footgear, called *"lapti,"* to become his nation's highest leader.

"I began working when I learned to walk," he used to say. "Till the age of 15 I tended calves, then sheep and then the landlord's cows. I did all that before I was 15. Then I worked at a factory owned by Germans, and later in a factory owned by a Frenchman. I worked at a Belgian-owned chemical plant, and now I am the Prime Minister of the great Soviet state."[1]

It was not hard to understand that Khrushchev should be a passionate advocate of his country's socialist system. Through it he had risen to the top. And as both Communist Party chief and Prime Minister he regularly delivered articulate expositions of its superior qualities. Although he got into plenty of trouble for it, Khrushchev never really backed down from his ominous-sounding warning that Communism would "bury" capitalism. He explained that famous line again and again, trying to siphon off from it any implication that the "burying" would be done by anything other than a peaceful, evolutionary process. Whenever Khrushchev made a public appearance he could be counted on to produce a few pithy sayings, possibly a few superlative ones, very frequently along the line that socialist progress marches forward, relentlessly, inevitably, to the ultimate detriment of all those who do not embrace it. Khrushchev was the great Communist true believer.

Just as there was hardly ever any advance warning of whether Khrushchev would show up, so there were no ready hints as to what he might say on any given occasion. May 1962 was still early in "the space race," when the United States under President John F. Kennedy was publicly striving to equal and overtake the spectacular Soviet achievements. On May 24, U.S. astronaut Commander Malcolm Scott Carpenter had successfully completed America's second orbital mission, following the flight of astronaut John Glenn on February 20. Carpenter, like Glenn, had flown through three orbits, landed two hundred miles down-range from his designated landing area, and floated in the Atlantic Ocean recovery area for about three hours before he was picked up. He had had his difficulties, too, in the Aurora 7 capsule, which tended to overheat. Yet he had returned safely, and even if American spacecraft were not as large in the early days as the Vostok ships, there were indications that the United States was making forward strides.

Suddenly at Sokolniki Park on that broiling May morning, correspondents spotted the thick-necked security men who accompanied Khrushchev on his outings, and then the rotund, perspiring Prime Minister appeared almost as if from nowhere. He moved slowly with his Italian escorts to the reviewing stands, which had been decked with Soviet and Italian colors for the opening of the exhibition. After remarks by his Italian hosts, Khrushchev approached the podium to say his words. At one point in earlier years he had threatened his own kind of "massive retaliation" against Italy should it become a launching pad for rocket attacks against the Soviet Union. On this occasion his line was pitched toward increasing trade and benefits through much closer commercial cooperation. This was of only minimal interest for American correspondents, but soon he steered around to the state of the cosmic art:

"Those who tried to liquidate trade between the capitalist countries and the Soviet Union," he exclaimed, his voice rising, "sat down in a puddle, while we took ourselves into the cosmos and we continue to have priority in the exploration of space. I congratulate Scott Carpenter because he really demonstrated great courage," Khrushchev went on, evidently ad-libbing. "If this courage had left him, he might have burned up in his spacecraft."

Western correspondents in Moscow are not always Soviet experts or highly fluent in Russian. Khrushchev was sometimes difficult to understand, particularly when he plunged into old Russian fables or convoluted rhetoric. On this occasion the only translation of the Prime Minister's remarks was into Italian. At least one news agency correspondent became confused as to just what Khrushchev had said, and telephoned an urgent message from the park to his office in Moscow for relay to New York—and it completely turned the sense upside down. The correspondent thought he had heard Khrushchev say that, because of the Carpenter flight, the Soviet Union no longer had priority in space. The erroneous report was widely circulated about the United States and the world. The dispatch reported in part:

"Now we are not alone in the cosmos," Khrushchev said. "Now the Americans have put two men in space." Mr. Khrushchev was understood by an Associated Press correspondent to say that the Soviet Union no longer has priority in space, now that the United States has put two men in orbit, just as had the Soviet Union.

The report was based on a misunderstanding and really did not square with Khrushchev's general attitudes about onward-and-upward Soviet progress. In journalism, though, sensational and inaccurate reports —such as a concession from Khrushchev that the Soviet Union had lost its lead in space—are very difficult to set right again. Subsequent corrections never seem to catch up. Khrushchev tried, however. That very same evening, at a Kremlin reception for the Italian industrialists, Khrushchev made a point of approaching the Western correspondents and chastising them for their sins. When the Communist Party newspaper *Pravda* published the transcript of Khrushchev's remarks, his phrase—"we continue to have priority in the exploration of space"—was there for all to read.

The incident illustrates how wedded Khrushchev was to the notion that the Soviet Union had surged ahead in space exploration—as indeed it had between 1957 and 1962—and how unthinkable it was to suggest that it might ever fall behind. Khrushchev and the Russian people were intensely proud of the achievements of Soviet scientists who had been working long years with little foreign input into their program of exploring space. Khrushchev, passionate advocate of Communism's superiority, liked to lord it over the Americans, and there were times when he seemingly engineered the quests of his scientists to coincide with his own political designs. The moon flight of September 1959, on the eve of his trip to the United States, was an example.

"The Americans have wanted so many times to send

their rockets to the moon, but without success," he would say on occasion, "and there is nothing to be done. They tell everyone they are launching rockets at the moon, but they go past. As we say sometimes, pulling the leg of a hunter, 'Bang, bang, you missed.' Thus with the Americans! But we go 'Bang, bang, and straight to the moon.' "[2]

Khrushchev died in September 1971. After his removal from power in October 1964, he lived in official disgrace and oblivion in a comfortable country house fifteen miles outside of Moscow. It is a shame that no curious visitor ever sat down and talked with him about Soviet goals and motivations in space either while he was in power or afterward. The memoirs attributed to Khrushchev, which were published in the United States at the end of 1970, throw practically no light on his intense preoccupation with cosmic exploration or his relations with his space scientists; yet Khrushchev surely held the answers to many fascinating questions about what went on behind the scenes. Nevertheless, one can gain a feeling for his general attitudes by rereading his speeches and picking out those comments which pertain to the space program. These could be arranged in the form of an interview, which might go something like this:

Question: Mr. Khrushchev, what was the purpose of launching the first two artificial satellites?

Khrushchev: The making and launching of the artificial sputniks ushered in a new era in scientific and technological development. The satellites will tremendously enrich our knowledge of the earth, its atmosphere, and outer space. Scientists are convinced that people will be able to embark upon interplanetary travel in the foreseeable future.[3]

Question: When do you contemplate throwing a man to the moon?

Khrushchev: You used a rather unfortunate phrase when you said "throwing" a man. We are not going to throw a man, because we value man highly and will not throw anyone. We will send a man into outer space when appropriate technical conditions have been developed. There are still no such conditions on hand. We don't intend "to throw" anyone in the sense, so to say, of throwing him overboard. We value people.[4]

Question: How well prepared are you to keep up the launching of artificial satellites after Sputnik-1 and Sputnik-2?

Khrushchev: The fact that we were able to launch the first sputnik and then, a month later, launch a second shows that we can launch ten, even twenty, satellites tomorrow. The satellite is the intercontinental ballistic missile with a different warhead. We change that warhead from a bomb to a scientific instrument and we launch a satellite.[5]

Question: The United States has relied to some extent on German experts in its missile development. What role did German scientists, who came to the Soviet Union at the end of the Second World War, play in developing your intercontinental missile?

Khrushchev: It is no secret that a small group of Germans did work in our country for a time and, on the expiration of their contracts, have either returned, or are returning, to Germany. When they returned and told what they knew, the Americans believed that they had reliable information about the stage reached by the Soviet Union in rocket building. We launched an artificial earth satellite, and the Americans complained afresh:

"We have been fooled again. The Germans who came to us know nothing about what the Russians are doing. It turns out that the Germans did not take part in developing the rocket."[6]

Question: Obviously, you have surrounded your rocket building program in secrecy and kept the German specialists at arm's length. Your achievements are truly impressive. Don't your scientists complain of lack of public recognition for their great efforts?

Khrushchev: The Soviet atomic experts and specialists who created the intercontinental rocket and sputniks have no complaint against their socialist country. They live so well that, God grant you a life like theirs, as the saying goes. The Soviet government has rewarded them, and many of them have received Lenin Prizes and the title of Hero of Socialist Labor. They are also well provided for from the material point of view. They "suffer" somewhat only in one respect—they are as yet anonymous as far as the outside world is concerned. They live, as it were, under the general designation of "scientists and engineers working on atomics and rocketry." But so far it is not widely known exactly who these people are. We shall erect a monument in honor of those who have created the rocket and sputniks and shall inscribe their glorious names in letters of gold so that they will be known to future generations throughout the ages.[7]

Question: What can you say about the effect which the launching of Soviet satellites has had on the world balance of power?

Khrushchev: The launching of the Soviet sputniks above all demonstrates the outstanding success achieved by the Soviet Union in the development of science and technology, and also the fact that the U.S.S.R. has outstripped the leading capitalist country—the United States—in the field of scientific and technical progress.

The launching of the sputniks undoubtedly also shows a change in favor of the socialist states that has been taken in the balance of forces between the socialist and capitalist states.

Balance of forces is a broad concept that includes political, economic, and military factors. The Soviet Union and the other socialist states are consistently pursuing a policy of peace and call for peaceful coexistence of states with differing social systems, for ending the arms race that is leading

toward a new war, and the prohibition of the use, production, and testing of atomic and hydrogen weapons.[8]

Question: Why did the Soviet Union consider it necessary
to proceed rapidly with the construction of the long-range
missile that you have used as your basic space booster?

Khrushchev: The American imperialists began to conduct a
nuclear policy of intimidating the socialist countries and
most of all the Soviet Union. Those were difficult times for
us. But what did our Party do? What did our people do?
They called on the best minds, physicists, mechanics, chem-
ists, and mathematicians, specialists of other branches of
sciences and technology, and put them to work. . . . Thus,
our scientists engaged in intensive work in their laboratories.
They thought, they calculated, experimented, and achieved
their own: they split the atom, and thereby showed Americans how talented are the Soviet people, what heights in the
development of science and in the practical application of
scientific knowledge they have achieved.

The American imperialists, having created the atomic
bomb, began to speak about the creation of the hydrogen
bomb. Our scientists took this into consideration, and
worked some more and created the hydrogen bomb earlier
than it was created in the United States.

The American imperialists surrounded our country with
their military bases and began to threaten us with these
bases. But what did our Party do, our government, our
scientists, engineers, and workers? Our scientists and designers, in response to the creation of bases around the
Soviet Union, created the intercontinental ballistic missile
in their laboratories, design bureaus, and factories. And all
the advantages which the American imperialists had possessed, having created military bases around our country,
they lost in the moment when our rocket took off and,
flying thousands of kilometers, accurately hit the region
designated by our scientists, engineers, and workers.[9]

Question: Do you think that the United States will also

develop an intercontinental ballistic missile that will be a threat to you?

Khrushchev: I can tell you that at the same time we told our scientists and engineers to develop intercontinental ballistic rockets, we told another group to work out means to combat such rockets.[10]

We expressed our great satisfaction with the work of the experts who produced the intercontinental rockets. At the same time, we remain very satisfied with the work of those who produced means for combating such rockets.

Question: Do you think you will be able to stay ahead of the United States in rocket development?

Khrushchev: The United States has set itself the task of catching up to the Soviet Union in the production of rockets in the course of the next five years [1960–4]. They will, naturally, join all forces to move rocket technology from the state in which it now finds itself and reach a more favorable position. But it would be naive to think that during this time we will sit on our hands. Indeed, even in the United States people are saying: And what are the Russians going to be doing—are they going just to pitch marbles?

Yes, naturally, we will do everything to take advantage of the time we have won in the development of rocket armaments and occupy the leading position in this area until such time as an international disarmament agreement is reached.[11]

Khrushchev's remarks sound quite reasonable. The Soviet Union, according to the Communist Party leader, was obliged to develop long-range rockets in order to defend itself adequately from the U.S. threat in the postwar years. These rockets were subsequently adapted for space exploration purposes before the United States was able to equal them. The Soviet Union intends to keep ahead of the United States in rocket development. Khrushchev's comments demonstrate a strong competitive posture, but also a characteristic caution about sending men off into space, and to the moon in particular. Certainly Khrushchev's competitive attitude and the open record of Soviet achievement were an essential part of "the space race," but they were only one set of ingredients. In the United States, President Kennedy also keenly articulated the essential competition of the two superpowers.

"The space race" was a symptom of superpower rivalry. Yet it was a competition curious in one sense. It was a race which was never really declared in an official way except by the Western press. Khrushchev affected a fighting stance. Yet he never pretended that Soviet space exploration was undertaken purely as a race or mammoth propaganda stunt. In the United States, the administration of President Eisenhower had consciously resisted the notion that the United States should compete for spectacular space achievements following Sputnik. Kennedy's attitude was more aggressive and came much closer to being an invitation to the race. To Kennedy, the Soviet satellites and the lack of comparable American spectaculars symbolized a lethargy that hit the United States under the Republican Party rule. And on May 25, 1961 (the year before Khrushchev's appearance in Sokolniki Park), Kennedy had gone to the Congress to make a historic statement in quite precise terms about American goals:

"I believe that this nation should commit itself to achieve the goal, before this decade is out, of landing a man on the moon and returning him safely to earth. No single space project in this period will be so difficult or expensive to accomplish.

"Let it be clear that I am asking the Congress and

the country to accept a firm commitment to a new course of action, a course which will last for many years and carry very heavy costs . . . if we are to go only half way, or reduce our sights in the face of difficulty, in my judgment it would be better not to go at all.

"This decision means a degree of dedication, organization and discipline which have not always characterized our research and development efforts.

"New objectives and new money cannot solve these problems. They could, in fact, aggravate them further unless every scientist, every engineer, every serviceman, every technician, contractor, and civil servant gives his personal pledge that this nation will move forward with the full speed of freedom, in the exciting adventure of space."

Kennedy was speaking just six weeks after the Soviet Union had orbited cosmonaut Yuri A. Gagarin, the first human being to travel into space. Kennedy's words were addressed primarily to the Congress and the American people, but when read in Moscow they must have represented the closest thing to a direct challenge to race to the moon. The Soviet leaders must have given some thought in their highest councils about the need for a response in words or in deeds. After the end of May 1961, there was an inherent compulsion on the part of the American press and other observers to discover whether Khrushchev was actively taking up the challenge. Within Congress, which would debate goals and appropriations, there was an intense interest to know the Soviet Union's purposes.

The United States had gone through a tortured recent history to arrive at this point—a history which differed

importantly from that of the Soviet Union. At the end
of the Second World War, the dominant desire in the
United States argued for a quick return to normal,
civilian life. In the Soviet Union vast destruction and
the conviction that the United States represented an
unpredictable military threat called for an austere program of reconstruction.

In 1946, American industry was unscathed by hostilities. America's economic potential remained great. Its
military capacity was unsurpassed. The United States
possessed air-atomic superiority, which included atomic
weapons whose destructive capacities had been demonstrated—possibly to excess—at Nagasaki and Hiro-
shima. And the United States operated a fleet of
long-range B-29 bombers with which to deliver the
charge. There was little incentive for President Truman
and his top advisers to plunge the nation into an expensive program of building intercontinental rockets.

Not only was the military need for such rockets questionable in the American estimation of the strategic
situation, but there was skepticism among U.S. experts
about the feasibility of building the entirely new weapons. How to construct a thin-walled system which would
be composed of nearly 90 per cent fuel? How to guide
such a vehicle over a flight of five thousand miles or
more? How to produce a warhead which would be
compact and less bulky than the original atomic bombs
dropped over the Japanese cities? How to solve the
problem of superheating on re-entry into the atmosphere,
which could quickly destroy a warhead? To many, such
problems seemed close to unsolvable. One former wartime science adviser, Vannevar Bush, is remembered
for his eloquence on the impracticality of the intercon-

tinental missile: "Its cost would be astronomical. As a means of carrying high explosive or any substitute thereof, it is a fantastical proposal. It would never stand the test of cost analysis. If we employed it in quantity, we would be economically exhausted long before the enemy."[12] Soviet scientists faced similar problems, but the demands of their governments and the sweep of their own imaginations were bolder.

There were warnings for the United States to observe in the years after the war, but American policymakers seemed mostly unmoved. The explosion of the Soviet atomic bomb, in 1949, should have clearly signaled a dynamic defense industry in the Soviet Union. The sophistication of the MiG-15 jet fighter in the Korean War, and the development of a hydrogen warhead in 1953 were also telltale signs. Additionally, in the early 1950s the United States began operating an electronic monitoring system from Turkey and elsewhere to track Soviet vertical rocket probes, and intermediate-range rocket tests. Returning German scientists contributed some knowledge of developments, but mostly their reports seem to have been misleading because of the isolated conditions in which they had lived and worked in Russia. The net result was that, by comparison, the United States government was dawdling.

Perhaps the United States government dawdled, but the military services, nonetheless, recognized rocket potential. They continued their basic research, although with considerable interservice rivalry and little coordination. The U.S. Army employed former German scientists, including the leading designer Dr. Wernher von Braun, under Operation Paperclip, beginning in 1945, to develop air-breathing winged missiles; the U.S. Navy concentrated on shipboard rockets which could be fired above and below the surface. The first U.S. intercontinental missile, the Atlas, was to emerge years later from research on a project called MX774, begun by the Consolidated Vultee Aviation Company on contract with the then Army Air Corps.

Just as the United States delayed over the development of long-range missiles, so too it delayed on development of a scientific orbiting satellite. The idea of such a satellite had been discussed in scientific circles, and the Army and Navy had both proposed satellite projects as early as 1946. The Rand Corporation, under contract to the Army Air Corps, even forecast the startling implications that would accrue to the nation orbiting a satellite first. The Rand study called the scientific satellite "one of the most potent scientific tools of the 20th century. Achievement of a satellite craft would inflame the imagination of mankind, and would probably produce repercussions in the world comparable to the explosion of the atomic bomb."[13] It was not until Western scientists created the idea of an International Geophysical Year that the satellite project caught on with top government echelons.

This, then, was the general historic background in the United States against which Sputnik went into orbit. The Central Intelligence Agency had warned the administration in 1955 that the Soviet 'Union was well on its way to accomplishing the feat, and by 1957 U.S. scientists were working feverishly. The Eisenhower Administration had some advance warning, but the general public was taken by surprise. The moment produced temporary confusion, a gamut of reactions. There were quick, slick comments from some administration officials who sought to downgrade the achievement. There were outraged exclamations from many legislators on Capitol Hill. But there began to occur a phenomenon that would continue for years, after each spectacular Soviet launch: a ground swell of opinion demanding that the United States overtake.

In the face of criticism, the Eisenhower Administration was at first defensive. The President saw no value in a race with the Soviet Union for the sake of showing off. "We never thought of our satellite program as one which was a race with the Soviets," declared White House Press Secretary James Hagerty. "Ours is geared to the I.G.Y. and is proceeding satisfactorily in accordance with scientific objectives."

But Sputnik, inevitably, ushered in an era of rethink. "The technological Pearl Harbor" caused people and legislators to question if the Soviet achievement had weakened the American defense posture. Had Sputnik proved American education wanting? Had it revealed basic flaws in American technological development? Had it disclosed that the United States generally underestimated the Soviet Union and failed to give it the attention it deserved? Serious expressions of national concern—that lasted for months—were the Senate hearings on rocket and satellite development conducted by Senator Lyndon B. Johnson at the end of 1957 and into 1958, as well as the series of television talks that President Eisenhower held in October and November 1957 to restore a sense of public calm.

The United States is a society where legislators and the press can raise a commotion, given an adequate cause. And in the days following Sputnik the public clamor became nearly irresistible. The administration reluctantly revised its space exploration program, and on April 2, 1958—six months after Sputnik—the White House sent to Congress draft legislation creating a civilian space agency. Under the administration of its first director, Keith Glennan, the National Aeronautics and Space Administration went on to develop both a short-term program and a longer-range one. On December 17, 1958, NASA did what the Soviet Goskommissiya never does, and announced the Mercury Program for putting a man into orbit around the earth. Toward the end of the Eisenhower Administration, the outlines of a ten-year program began to take shape; it included the development of the huge Saturn moon rocket, which, it was thought, could take men to the lunar surface and back sometime in the decade *after* 1970. But there was still an aura of indecision about these farseeing visions: "Further testing and experimentation will be necessary to determine whether there are any valid scientific grounds for extending manned flight beyond the Mercury Program," Eisenhower cautioned in his final budget message to Congress before turning over the Presidency to Kennedy.

Kennedy had made much of the space lag during the election campaign of 1960. However, his views on U.S. priorities in space were still in process of formulation when he took office in January 1961. Kennedy quickly began to re-examine the American program and during his first one hundred days asked the NASA administrator, James E. Webb, to offer a revised budget. Webb proposed a modest expansion of $308 million, which was initially turned down by the Bureau of the Budget. As late as March 1961, Kennedy reportedly was still asking questions about the financing and construction of the Apollo spacecraft.

Kennedy had considerable advance notice about Soviet capability and intention to launch a man into orbit around the earth. Between May 1960 and March 1961, the Russians released details of five flights, some carrying dummy cosmonauts, which were quite clearly precursors to a manned flight. Furthermore, U.S. electronic intelligence provided the Chief Executive with a tipoff that final preparations on the Baikonur launching pad were in train. Kennedy, reportedly, drafted a message of congratulations to the Soviet Union that he held in abeyance until the Vostok rocket with Gagarin on board took off from the Central Asian Cosmodrome. Nevertheless, the flight of April 12, 1961, made a deep impression on the President. Two days later he summoned a group of advisers to the White House to discuss how the United States might counter Soviet achievements. Kennedy asked Vice President Johnson to conduct an investigation, and posed a number of quite specific questions to which he desired answers:

1. Do we have a chance of beating the Soviets by putting a laboratory into space, or by a trip around the moon, or by a rocket to go to the moon and back with a man? Is there any other space program which promises dramatic results in which we could win?

2. How much additional would it cost?

3. Are we working 24 hours a day on existing programs? If not, why not? If not, will you make recommendations to me as to how work can be speeded up?

4. In building a large booster, should we put our emphasis on nuclear, chemical or liquid fuel, or a combination of both?

5. Are we making maximum efforts? Are we achieving necessary results?[14]

The results of the investigation produced difficult answers for Kennedy. There was no early, dramatic counterachievement which the United States could produce, given its current stage of rocket development. The most promising hope for overtaking the Russians, it seemed, would be a manned lunar mission. This might be accomplished with a safe return to earth no earlier than 1967 or 1968. NASA administrator Webb viewed a lunar mission with a general deadline as a useful goal for harnessing U.S. industry and thereby giving a significant boost to the development of modern technology. Caution dictated, however, that the deadline for the mission should be phrased in a manner which would convey immediacy, and yet allow for slippage. Thus, the slightly vague formulation "before this decade is out."

The exact mix of motivations that finally convinced Kennedy to accept this project may never be known with certainty. Scientific inquiry, the technological boost it would give to the United States, the expectable value of unforeseen spinoff were powerful reasons; also Kennedy's concern with "getting America going again." And the President's personal image was tarnished by the disastrous failure of the Bay of Pigs invasion in Cuba. In any event, on May 25, 1961, he appeared before the Congress and announced this goal. From then onwards, as it debated American space appropriations, the Congress would be looking for every indication that the Russians might be losing their enthusiasm for the space endeavor.

It is interesting to note that Khrushchev and other Soviet officials were well aware of the Kennedy challenge. They replied, primarily, that space was big enough for all nations to explore; that the Soviet Union was waiting for the United States to catch up and to equal Soviet achievements; that America would certainly have a lot of catching up to do before it ever became a serious challenger to the superiority of the Soviet Union. Occasionally, in the early 1960s, Soviet cosmonauts declared that when American astronauts landed on the moon, Russians would be already there to greet them. Such remarks tended to be interpreted in the United States as reflecting a secret Soviet timetable for beating the United States to the moon. However, it is notable in retrospect that neither Khrushchev nor any authoritative Soviet official like Korolyov ever directly took up the Kennedy challenge. Usually Soviet officials adopted a cautious, even evasive, attitude in describing plans for a manned lunar mission. Evidently Khrushchev had been well briefed on the difficulties of sending men to the moon and bringing them back safe and sound.

Beginning in the late 1950s, Khrushchev did, however, throw out an economic challenge for the year 1970—a challenge intended first for his domestic audience, but frequently elaborated for his foreign listeners as well.

A long history of predictions by Soviet leaders goes back to Lenin, who insisted that the Soviet Union would eventually overtake the capitalist West in industrial and agricultural output. In 1919, the Bolshevik leader declared: "Either perish or overtake the advanced countries." In 1931, Stalin added: "We are fifty to a hundred years behind the advanced countries. We must make good this distance in ten years. Either we do it, or they will crush us." Khrushchev added his own formulations following the 20th Communist Party Congress of 1956, which revised Communist ideology to enshrine the concept of "peaceful coexistence" of states with differing social systems. Then, about 1957, Khrushchev began to focus on the economic race between the United States and Soviet Union; he did not hesitate to declare the Soviet intention to compete or reveal its goals:

"Some people in foreign countries used to ask, 'Mr. Khrushchev, do you really expect to catch up with America economically?' Today, no one raises this question in this way. Instead, I am asked, 'Mr. Khrushchev, what do you think? In what year will you catch up with America?' They no longer doubt that the Soviet Union will overtake the U.S.A. Now they are troubled by one question—when? My reply is, 'You can write in your little notebook that we will overtake you in per capita industrial production by 1970.' "[15]

It was a bold prediction. Khrushchev went much further still, apparently with the cooperation of his most optimistic economists. After consideration for nearly four years, Khrushchev produced in July 1961 the draft for the Third Communist Party Program to be adopted by the 22nd Congress of the Communist Party that fall. The program laid out in some detail the goals for increasing Soviet industrial production in numerous sectors of the economy. The program was adopted in October 1961 to the skepticism of Western economic observers. As the years passed, the skeptics were proved right. The goals had to be successively downgraded in 1966, 1967, and 1969. Well before the end of the decade, Soviet authorities stopped talking about the ambitious Party Program and the economic race against the United States. The words of the 1961 Soviet economic challenge, nevertheless, contrasted interestingly with President Kennedy's appeal:

"The Party sets the task of converting our country, within the next decade, into the world's leading industrial power, and of winning preponderance over the United States both in aggregate industrial output and in industrial output per head of population. By approximately the same time, the U.S.S.R. will exceed the present level of U.S. agricultural output per head of population by 50 per cent, and will surpass the U.S. level of national income.

"But that is only the first objective. We shall not stop at that. In the course of the second decade, by 1980, our country will leave the United States far behind in industrial and agricultural output per head of population."[16]

Both Kennedy and Khrushchev tried to harness the superpower rivalry between the United States and the Soviet Union for the purpose of stimulating their own domestic programs—in the American case, space exploration; in the Soviet, economic development. Neither thought it worthwhile directly to respond to the other's challenge.

Khrushchev was questioned directly from time to time about the Soviet Union's intention to fly men to the moon in view of the American intention to do so. His answers to such questions were usually cautious and capable of varying interpretations. One response was especially memorable. It occurred at a Kremlin press conference October 26, 1963, which had been convened especially for a visiting group of left-wing journalists from developing countries. Resident foreign correspondents in Moscow were excluded, but the transcript of Khrushchev's remarks was quickly made available in the government newspaper *Izvestia*, and his comments caused a sensation. Khrushchev was asked by a journalist from Latin America about a Soviet manned moon flight, and he replied in these terms:

"It would be very interesting to take a trip to the moon. But I cannot say when it will be done. We are not at present planning flights by cosmonauts to the moon. Soviet scientists are working on this problem. It is being studied as a scientific problem and the necessary research is being done. I have a report to the effect that the Americans want to land a man on the moon by 1970–80. Well, let's wish them success. We shall see how they fly there, and how they will land on the moon, and more important, how they will start off and fly home. We shall take their experience into account. We do not want to compete in the sending of people to the moon without careful preparation. It is clear that no benefits would be derived from such a competition. On the contrary, it would be harmful as it might result in the destruction of people. We have a frequently quoted joke: he who cannot bear it any longer on earth may fly to the moon. But we are all right on earth; to speak seriously, much work will have to be done and good preparations made for a successful flight to the moon by man."

The next day newspapers all over the world headlined the news: KHRUSHCHEV DROPS OUT OF THE MOON RACE. It was an oversimplification of what he had said, and Khrushchev objected strenuously. A little over a week later, the First Secretary seized an opportunity to clarify his press conference performance. The occasion was a trip to Moscow by a group of American business executives. They arrived in the Soviet capital at a time when considerable attention was focused on the Soviet economy. 1963 had been a disastrous year for agricultural output, and the U.S.S.R. was forced to convert an enormous amount of gold to pay for imports of wheat from the United States, Australia, and other countries. Inevitably, in their private conversations, the business-men asked Khrushchev about the economic resilience of the Soviet Union and Russian plans in the costly, and possibly unnecessary, exploration of space.

Khrushchev parried any suggestion that his country's economic underpinnings were tottering and might force a drastic cutback in space activities. "Gentlemen, give up such hopes once and for all and just throw them away," he told the visitors. "When we have the technical possibilities of doing this [sending a manned expedition to the moon] and when we have complete confidence that whoever is sent to the moon can safely be sent back, then it is quite feasible. We never said we are giving up our lunar project. You are the ones who said that."[17]

What had happened to cause this confusion was another example of the frailties of the press, and this author was directly at fault for oversimplifying Khrush-chev's original remarks. The edition of *Izvestia* con-taining the Khrushchev transcript became available late on Saturday night, October 26. Seeking to catch as many deadlines of Sunday newspapers as possible, I hurriedly transmitted the moon flight comments without ade-quately supplying the qualifying remarks. Khrushchev was essentially rephrasing Korolyov's 1963 explanation that much work needed to be done before a manned trip to the moon would be feasible and that work would take more than one year. The Communist leader did not categorically rule out a flight to the moon although he did seem to reject the notion that a moon flight should be in direct competition with the United States.

chapter 6

CONCEDING
THE MOON RACE

On March 18, 1965, the Soviet Union made a spectacular surge forward in "the space race." The two-man
Voskhod-2 spaceship sailed into orbit and, through a
collapsible airlock, cosmonaut Alexei A. Leonov wrig-
gled into the cosmic void to become the first human
being to cavort in space. Within three months, the
United States orbited Gemini 4 and Major Edward H.
White II followed and surpassed Leonov's feat with a
lengthier, more complicated performance of space walk-
ing. The year 1965 was the pivotal point in the observ-
able competition of spectaculars. In the months that
followed, both nations continued broad programs of
space exploration highlighted by pioneering feats. How-

ever, just over three years later, in 1968, the Soviet Union began to concede in a quiet and restrained manner that it was behind in the effort to send a manned expedition to the moon.

For a long time it continued to look as if the Soviet Union was in the race. In September 1968, Soviet scientists sent Zond-5, an unmanned automatic probe, around the moon and brought it back to a splashdown in the Indian Ocean. This feat was also a "first." It was followed in November by the similar journey of Zond-6. Soviet scientists openly hinted that the Zonds could have carried men. And officials at the National Aeronautics and Space Administration strongly suspected that the Kremlin was seeking to outdo the United States by executing the first manned circumlunar flight. In Washington, an impressive counterproject was approved for the American Apollo moon program at the end of 1968. And, accordingly, Apollo 8, with a three-man crew commanded by Colonel Frank Borman, was launched on a Christmas flight around the moon. It returned to a successful landing in the Pacific Ocean. American astronauts had accomplished the voyage which Jules Verne described nearly a century before and so many other dreamers had contemplated ever since.

Soviet scientists could plainly see that the United States was serious about its moon project, and they perceived the likely course of events. In numerous statements, frequently overlooked by the American press, the Russians acknowledged that they would have liked to have been first around the moon with men. They added that the Soviet Union had decided on another course: it would explore the universe with automated devices.[1]

The commentaries were so subdued that it was difficult to be sure of Soviet intentions. The remarks could have been a verbal diversion, aimed at dissipating attention while some secret moon project forged ahead. With the passage of time, events have become clearer, although the Kremlin has yet to give a comprehensive view of its decision to abandon a manned moon expedition.

To the Russians, a manned moon mission was, in the 1960s, the greatest of the achievable spectaculars of the near future. It carried with it rich rewards of prestige, but its costs were high. It required large expenditures, a long-term commitment, extremely high technological standards, the best in management techniques, and, of course, boundless human courage. The Soviet Union had these resources in varying degrees, and their presence in Russia has become obvious to both casual observer and space expert. In 1970, Russian scientists demonstrated an engineering triumph with two unmanned lunar flights. Luna-16 took off in September 1970 and soft-landed on the moon. There, by remote control, it scooped up a handful of moon dust (101 grams), packed it, stored it, and then blasted back to earth. Although the payload was minute in comparison with the hundreds of pounds of moon rocks returned by American astronauts, the Soviet experiment proved that unmanned exploration of the universe could become infinitely sophisticated. Luna-17 added a further dimension. This probe, after soft-landing on the moon in November 1970, released an eight-wheeled moon rover, called Lunakhod-1, which traveled more than a mile from the parent ship and survived about ten months. While cruising over the lunar surface, Lunakhod-1 carried out a number of experiments, snapped a series of photographs, and radioed its data back to earth.

The Soviet program was never directed solely toward exploring the moon, and, consequently, the abandonment of the manned moon project did not leave a gap in Russian plans. Soviet scientists, for example, expended considerable energy in probing Venus, and to a lesser extent Mars. On May 19 and May 28, 1971, they launched Mars-2 and Mars-3 after making a number of unsuccessful, and unnamed, tries in previous years, according to Western experts. The Russians have launched vehicles toward Venus at almost every opportunity since 1961. During 1970, the Russians accomplished another "first" by parachuting an automatic probe onto the Venusian surface. Venera-7 survived for twenty-three minutes under intense heat and stupefying pressures, transmitting basic data about surface conditions. Another Venus probe, which would have been labeled Venera-8, failed to leave its parking orbit around the earth and decayed after seventy-six days. Soviet authorities gave it the name of "Kosmos-359."[2]

In addition to the unmanned probes, the Russians have continued an active program of manned orbital flights. The Russian program, in contrast to the Apollo moon program, has put emphasis on the creation of a manned laboratory around the earth. This goal was repeatedly described by Korolyov. He envisaged both an orbiting laboratory and, eventually, a space platform that could be used as a staging area for manned expeditions to the moon and planets.

The group flights in 1962 and 1964 of his Vostok single-seater craft were early precursors that explored the problems of docking spacecraft in orbit. These ships approached each other in orbit but did not link up. For that experiment, Korolyov and his colleagues created the Soyuz ship, a much larger vehicle which could accommodate three men and stay in orbit up to a month. The flight of Soyuz-1 in April 1967 ended tragically when the parachute lines became entangled following re-entry and cosmonaut Vladimir M. Komarov was killed. But the ship was perfected, and from 1968 through 1971 carried out a continuing series of experiments. In 1969, two Soyuz craft joined, transferred crews (by the relatively unsophisticated method of crawling outside the ships), and demonstrated that a primitive earth orbital station was possible.

Finally, on June 7, 1971, Soviet scientists began a project which achieved Korolyov's long-standing goal of creating an orbital space station. A module called "Salyut" was launched into orbit and joined by a three-man spaceship, Soyuz-11. The Salyut-Soyuz combination measured some sixty feet in length, weighed about twenty-five metric tons, and afforded some one hundred cubic meters of space. It appeared that the Salyut module was so constructed that it could serve as a "hub" and would be able to dock with visiting spacecraft.

The flight of the Soyuz-11 ship, however, ended tragically on June 30, 1971. The ship was observed to make a perfect re-entry and landing. But when recovery teams unbolted the exit hatch they found cosmonauts Lieutenant Colonel Georgy T. Dobrovolsky, Vladislav N. Volkov, and Viktor I. Patsayev to be dead. The Soviet government, plunged into mourning, immediately appointed a special commission to investigate the catastrophe. Non-official Soviet sources shortly reported that during the final re-entry phase the Soyuz cabin had depressurized and that the cosmonauts died of oxygen loss and air bubbles forming in their blood stream. Almost simultaneously, the Soviet press published an article by academician Boris N. Petrov, a prominent space manager, asserting that the Russian commitment to orbital stations was as constant and solid as ever.

On other fronts, Soviet military and civilian scientists made energetic use of space technology. By mid-1971, more than four hundred scientific satellites of the "Kosmos" series had been launched. These satellites, and others, have been used to develop weather prediction, improve navigation, map inaccessible regions, look for earth resources, and improve communications. The Soviet Union has improved telephone and television links through the use of the "Orbita" system, which employs Molniya satellites, visible across the five-thousand-mile expanse of the Soviet Union. The Soviet military has used space technology (and occasionally the cover of the Kosmos series) to introduce reconnaissance satellites, electronic ferret satellites, and communication, navigation, and weather satellites. Military scientists have developed a bomb which can be launched from orbit (the so-called fractional orbital bombardment system), as well as an "inspector-destructor" satellite. Western experts report that this latter device, still in an experimental stage, should be able to demolish a cosmic intruder.

If the known civilian and military launches of the Soviet Union and United States are analyzed and compared, it is clear that both countries are making concentrated efforts in space. Between 1957 and 1970, according to one tabulation, the United States launched 241 civil-oriented craft to 171 for the Soviet Union; the

United States also launched 273 military craft to 294 for the Soviet Union.[3]

How much the Soviet Union spends on its space activity is not precisely known. Soviet security precautions prohibit the release of budget information on the space program. The annual Soviet government budgets presented at the end of the calendar year to the Supreme Soviet are not highly informative. These budgets present an overall figure for scientific research, but the amount is not broken down into specific categories. Western experts have studied Soviet space activity and tried to calculate its cost. According to such estimates the Soviet Union is believed to spend as much as 2 per cent of its gross national product on space. Since the Soviet GNP is rated at about $500 billion a year, Soviet space expenditures might be as much as $10 billion annually. That is a sizeable, and impressive, figure.[4]

When Western observers have suggested that this sum is more than the Soviet population—and economy —can comfortably afford, Soviet leaders have rejected the notion. Premier Aleksei N. Kosygin asserted in December 1965 that space exploration was well within the reach of the Soviet economy: "I would say that man will always go on seeking a solution to the problem of the universe. There will always be funds which will be set aside to resolve the problems relating to the world and the universe; this is all to the good if it is purely scientific. We don't have any contradictions in the Soviet Union between appropriations for space research and for the needs of the population, or education and such. They are a negligible part of overall expenditure. Space expenses do not detract from the needs of the population."[5]

Against this background, the fate of the manned moon expedition should be viewed. To begin, then: Is it certain that the Soviet Union ever really committed itself to land men on the moon before the United States, or even at about the same time as the United States? This is a worthwhile question. If the Soviet Union thought it useful to beat the United States to the moon, for whatever reasons, President Kennedy's determination to perform the feat by the end of the 1960s had a justification worthy of the most serious consideration. On the other hand, if the Kremlin had dropped its plans for an early manned lunar landing, there is reason to question the costs of the American commitment. As history turned out, there was minimal communication between Washington and Moscow on the manned lunar goal. Did the lack of communication and cooperation lead the United States to spend vast amounts of money on the Apollo moon landing when it had other, competing necessities?

Over the years, the president of the Soviet Academy of Sciences, Mstislav V. Keldysh, has been asked repeatedly and in many different ways whether his country intended to land men on the moon. In the fall of 1969, he answered that the Soviet Union had put aside such plans for the moment. In 1970, his answer was more enigmatic: "We never announced such a program, therefore there is nothing really to give up. In the near future, as I have said many times, we are not planning manned flights to the moon."[6]

It is true that the Soviet authorities never announced a manned lunar landing project as the United States did the Apollo program. And this is not surprising. It would go against Soviet practice to announce mission goals in advance. Yet, during the first half of the 1960s any number of Soviet figures—top political personalities, the Chief Designer Korolyov, the cosmonauts, space scientists—carried on a lively discussion about Soviet plans for landing men on the moon. Cosmonaut Gherman S. Titov said in 1962 that the Soviet Union was thinking in terms of a precursor, circumlunar flight about 1965. Cosmonaut Valentina V. Tereshkova said during a trip to Cuba in 1963 that Yuri A. Gagarin had been selected to lead the first moon expedition. Cosmonaut Valery F. Bykovsky described in early 1968 how a manned flight would be preceded by an automatic craft whose passengers would be animals. Scientists carried on a discussion in the Soviet press on the best way to reach the moon, the majority seeming to favor an expedition assembled in space and departing from an orbiting platform. A U.S. government study has collected these various statements between 1961 and 1964. The inescapable conclusion is that at least until Khrushchev's ouster in October 1964 the Soviet Union was intent on landing men on the moon and was seriously studying how to do it.[7]

The Soviet space watcher can observe a general pattern in Soviet statements about the moon landing goal. From Sputnik to the fall of 1968 there was a ring of certainty that the first men on the moon would be Russians and that they would be there to greet Americans. The fall of 1968 is the turning point for such statements, with Titov still insisting in Mexico City on October 23 that the Russians would be first.[8] As 1968 drew to a close (with the highly successful Apollo 8 flight) the emphasis changed. Soviet scientists began stressing the advantages of unmanned exploration. It was, they said, far less costly—possibly twenty to twenty-five times less costly—than manned exploration. And an unmanned exploration eliminated the dangers to man. It looked, on the surface, as if the Kremlin had made a decision between 1964 and 1968 against the lunar expedition. This hypothesis can be tested by examining various, intriguing bits of evidence that have surfaced, and by looking at the space hardware the Soviet Union actually developed.

The first striking point is that there were important differences between Khrushchev and his space scientists. Khrushchev, as evidenced by his statements, was a partisan of the manned lunar landing when it could be accomplished with reliable guarantees for the cosmonauts' safety. But there were Soviet scientists who were skeptical of the costs and the utility of sending men to the moon. These skeptics have not yet been identified, but their existence has been alluded to in the Soviet daily press as well as in a number of space publications. Some of the most explicit evidence of the skeptics' existence and influence was contained in a letter written by Sir Bernard Lovell, director of the Jodrell Bank radio telescope near Manchester, England.

Sir Bernard had helped the Russians in tracking a number of their early moon shots. In gratitude, he was invited to visit Russia in 1963 and traveled over twelve thousand miles during a three-week visit. On his return, he reported that he had been to a number of installations that Western visitors had never seen before, including the deep-space tracking center in the Crimea. In his letter of July 23, 1963, to the deputy administrator of the

National Aeronautics and Space Administration, Dr. Hugh Dryden, Sir Bernard reported that the Soviet Academy of Sciences was questioning the value of a manned lunar landing and favored its indefinite postponement. The Soviet aims, according to Sir Bernard, were:

(A) A determination to perfect the rendezvous technique with an immediate aim (perhaps 1965–1966) of establishing a manned space platform for astronomical observations at a height of 150–200 miles. The duty period of the astronauts on this platform is envisaged as 5–7 days with immediate return to earth if lethal solar radiation seems probable.

(B) Continuation of the plans to implement the existing programs on the instrumental exploration of the moon, Venus and Mars. I think it can be assumed that apparatus is now in process of assembly for the attempt to make a soft landing of instruments on the lunar surface, and that the launching will be made in a matter of months.

(C) The rejection (at least for the time being) of the plans for the manned lunar landing. The President (Keldysh) gave three reasons:

(1) Soviet scientists could see no immediate solution to the problem of protecting the cosmonauts from the lethal effects of intense solar outbursts.

(2) No economically practical solution could be seen of launching sufficient material on the moon for a useful manned exercise with reasonable guarantee of safe return to earth.

(3) The Academy is convinced that the scientific problems involved in the lunar exploration can be solved more cheaply and quickly by their unmanned, instrumented lunar program.

In the subsequent discussion I told the President that, personally, I did not agree with this assessment since I believe that the human brain was essential to the efficient solution of the problems presented by the lunar surface. He replied that the manned project might be revived if progress in the next few years gave hope of a solution of their problems, and that he believed the appropriate procedure would be to formulate the task on an international basis. He stated that the Academy believed that the time was now appropriate for scientists to formulate on an international basis (a) the reasons why it is desirable to engage in the manned lunar enterprise, and (b) to draw up a list of scientific tasks which a man on the moon could deal with which could not be solved by instruments alone. The Academy regarded this initial step as the first and most vital in any plan for proceeding on an international basis.[9]

Sir Bernard further expounded these views in meetings and interviews with the press after his visit. Yet his reports were greeted skeptically, particularly in the United States. Officials at the National Aeronautics and Space Administration had developed tentative relations with Soviet space scientists beginning in 1962. They felt that if the Soviet government were serious about discussing the possibility of joint exploration of the moon with men it would communicate directly with Washington. Furthermore, Sir Bernard's reports seemed to be contradicted by the main thrust of Soviet public statements on the value of a manned moon landing and most notably Khrushchev's. Finally, in October 1963, at a press conference in Prague, Keldysh seemed to repudiate Sir Bernard directly. The president of the Soviet Academy of Sciences claimed he had not told the visiting English professor that the Soviet government had abandoned the manned moon landing, and asserted that this must be a conclusion which Sir Bernard had arrived at all by himself.[10]

In retrospect, Sir Bernard's report seems, nevertheless, particularly prescient. The Soviet Union has moved forward in unmanned exploration of the moon. It has carried out probes of Venus with automatic devices. It is actively working on a manned space platform. And no manned lunar mission has been mounted. Sir Bernard, commenting in 1970 on his July 1963 letter, states that he believes it represented "a reasonably accurate statement of what actually happened." He adds that the "main departures from actuality have been in the time scale." He does not take too seriously the Keldysh denial and remarks that his relations with the Soviet science administrator have continued to be "extremely friendly."[11]

Khrushchev, in any case, still seemed bent on continuing active studies of the lunar mission, and one can imagine that he had a variety of reasons for this. His motivations may have had points in common with prevailing views in the United States. By continuing a manned moon program, the Soviet Union would be obliged to organize its industry to confront the most advanced technological challenges. This effort would have spinoff benefits in many areas, some obvious as in the case of satellite communications, and some impossible to foresee. A manned landing on the moon ahead of the United States would prove, in political terms, a formidable prize with myriad ramifications. The Soviet Union would gain an enormous increase in prestige, might witness some favorable shifting in Communist opinion toward Moscow in the Sino-Soviet ideological dispute (which by 1963 was growing even sharper), and might win new interests in trade with European and other nations. Such considerations of high policy have yet to be disclosed by Soviet officials. Khrushchev, in the memoirs attributed to him in 1970, unfortunately did not touch upon the subject.

There is some evidence that Khrushchev caused tension among many scientists, including Korolyov, with his goal of putting a manned expedition on the moon. One account by a recent Soviet defector asserts that Khrushchev definitely tried to elicit sensational technological achievements from his scientists, occasionally over and above the interests of science. According to former Moscow journalist and editor Leonid Vladimirov, Khrushchev set a crude form of pressure against Korolyov by establishing a competing group of scientists led by academicians Mikhail K. Yangel and Vladimir N. Chelomey. Khrushchev's pressuring reportedly forced Korolyov to take drastic steps to develop a multi-manned spacecraft, the Voskhod. Korolyov adapted the Voskhod from the Vostok spaceship that first carried Gagarin aloft. The Chief Designer eliminated the catapult that served as a safety device during the final descent stage, and made room for up to three men. He added new equipment, including retro-rockets to assure a soft landing, and called the improved Vostok ship the Voskhod. Voskhod-1, carrying a three-man crew, was launched into orbit October 12, 1964, to be followed March 18, 1965, by Voskhod-2, which carried the world's first space walker. Vladimirov's account asserts that Korolyov achieved these modifications by cutting corners under Khrushchev's pressuring.[12]

Soviet officials regard such accounts as Vladimirov's with contempt. The officials contend that these émigré reports are aimed at blackening the Soviet image; and they shed no new light on the situation. Yet there have appeared, in officially approved Soviet publications, hints that Korolyov and Khrushchev sometimes found themselves at odds. One piece of evidence appeared in 1967 in a six-hundred-page history of Soviet aerospace activities. The volume, *Aviation and Cosmonautics U.S.S.R.*, credits Korolyov with being the driving force behind the Soviet government's broad program for conquering the cosmos. In a particularly significant but guarded passage, the history notes that Korolyov was "merciless toward unfounded fantasizing." The Russian word for fantasizing in this passage is *prozhektorstvo*—the very word which was leveled against Khrushchev by the new Soviet leaders after they ousted him from power in October 1964 and used repeatedly since then in the Soviet press to describe one of the former leader's unpardonable failings. One could, therefore, conclude that this passage is a veiled acknowledgment that Khrushchev sought to elicit from his space scientists achievements that were scientifically questionable and, possibly, undesirable from an economic point of view.[13]

In the end, Khrushchev fell from grace. He was removed by a carefully conceived plot in the first half of October 1964. Those who acted against him, including some of his closest colleagues on the Communist Party Presidium, were exasperated by his failure to consult regularly, his refusal to accept sound scientific and managerial advice regarding a long list of domestic problems, his increasing tendency to take unilateral decisions affecting Soviet policy, and his wild boasting for international public consumption.

Khrushchev's fall was intertwined with the space drama of Voskhod-1. On October 12, this ship lifted into earth orbit the first three-man crew in the world. Commanding the mission was Colonel Vladimir M.

Komarov. He was accompanied by Dr. Boris Y. Yegorov, a twenty-six-year-old doctor, and Konstantin P. Feoktistov, a close colleague of Korolyov's and later a prominent manager in the Soviet space program. The flight was not only the first of a Voskhod craft, but the first time scientists had been put in orbit to take varied, and careful, observations of man and his abilities in space.

Khrushchev had quietly left Moscow earlier in October for a vacation at his Black Sea retreat in Pitsunda. Shortly after the launch from the Baikonur Cosmodrome, he held a nationally televised conversation with the three-man crew. Bubbling with enthusiasm, he talked excitedly of the great welcome in Red Square he planned for the cosmonauts' return: "I warn you, so to say, that you managed quite well with the gravity overloads during takeoff, but be ready for the overloads which we will arrange for you after you come back to earth. Then we'll meet you in Moscow with all the honors you deserve."[14]

When Khrushchev's face disappeared from the television screen, it was forever. Immediately thereafter, the events which led to his removal began to unfold. In the meantime, he hurriedly managed to receive one last foreign visitor, the French minister of atomic energy Gaston Palewsky. Khrushchev abruptly canceled a planned lunch for the French visitor on October 13, received him for a brief half-hour chat in the morning, and then dashed back to Moscow to confront his accusers.

Simultaneously, the events developing secretly on the ground extended out into space to touch the mission of Voskhod-1. Western observers believed that Voskhod-1 was equipped to stay in orbit up to one week, but orders were dispatched by Korolyov to end the flight at the end of the first day. Soviet media recorded his enigmatic command, and published the transcript:

Korolyov: Are you ready to proceed to the completion of the final part of the program?

Komarov: The crew is ready. But we would like to prolong the flight.

Korolyov: I read you, but we had no such agreement.

Komarov: We've seen many interesting things. We would like to extend the observations.

Korolyov (joking): "There are more things in heaven and earth, Horatio . . ." We shall go, nevertheless, by the program.[15]

The passage which Korolyov was quoting from Shakespeare's *Hamlet* says bluntly enough: "There are more things in heaven and earth, Horatio,/Than are dreamt of in your philosophy." In this context it looks very much as if Korolyov was passing on a hint that something extraordinary was happening at home. Soviet sources have always contended that Voskhod-1 fully completed its mission, the implication being that the flight was not cut short. There seems a good possibility, however, that the flight was confined to a minimum program, and that Khrushchev's challengers called down the mission as a precaution. They did not want to be distracted if trouble developed in space.

A number of events occurred in the first several years of the new Soviet administration which had an effect on the development of the space program. It can be assumed that on Khrushchev's resignation, the new Communist Party chief, Leonid I. Brezhnev, and his colleagues in the Politburo (the name of this top Communist Party body was changed from Presidium after Khrushchev's ouster) undertook a thorough review of the major economic and political problems confronting the nation, including the space program. Vladimirov asserts that the new leaders heard a frank report by Korolyov on the progress in space being made by the United States. The former Soviet journalist says that it was decided, in view of the anticipated achievements by the United States, to cease propaganda about reaching the moon first.[16] How true this part of Vladimirov's report may be is still subject to further verification. There continued to be isolated claims by cosmonauts and others about Soviet plans for a moon landing that would take place before Apollo 11's anticipated touchdown at the Sea of Tranquillity.

Soviet authorities have not released any details about a post-Khrushchev reappraisal of the space program, but Brezhnev's sober attitude on "the space race" was quickly made known. At the belated Kremlin reception for the returning Voskhod-1 cosmonauts on October 21, 1964, the General Secretary of the Communist Party painted the Kremlin attitude on space in this way:

"We Soviet people do not look on our space exploration as an end in itself, as some sort of 'race.' The spirit of gamblers is profoundly alien to us in the great and serious business of exploring and conquering cosmic space. We regard this enterprise as a component of the tremendous, creative work in which the Soviet people is engaged, consistent with the general line of our party in all areas of the economy, science, and culture in the name of man and for the good of man."[17]

It is reasonable to conclude that the new Kremlin leadership eased off whatever pressures Khrushchev had brought to bear on the scientists to beat the United States to a manned moon landing. This does not mean, however, that Brezhnev and his colleagues quickly reached a decision to abandon the project. Indications are that the Kremlin leaders delayed a precise ruling while they waited to see how their space technology would develop in the coming months. Soviet scientists had under development a variety of equipment, including a giant booster that eventually would have the capability of taking a manned expedition to the moon, a space-walking experiment with the Voskhod-2 ship which Korolyov had fitted with a special airlock, and an automatic craft for landing an instrument package on the moon.

During the first two years of the new Kremlin administration the space program made several remarkable achievements. The first of these was Leonov's dramatic walk in space. Leonov had become the first man to venture into space while his ship hurtled around the globe at some 18,000 miles an hour. He proved what scientists had expected: that man can exist in a specially equipped spacesuit in outer space and perform a variety of tasks there although the process is an arduous one.

The success of Leonov's space walk was hailed at a later press conference in the marble halls of Moscow State University. At the time the feat seemed to be a new argument for the scientific potential of man in space. Observers concluded that the Soviet Union would continue experiments along this line, and that Leonov would eventually prove to be the precursor of a more dramatic walk in space—specifically across the lunar surface. But the second event at the end of January

1966 had, by logical extension, different implications. This was the first soft landing of an instrument package, Luna-9, on the moon's surface. Luna-9 was a comparatively primitive device whose soft-landing technique could not be used for a human crew. Unlike the U.S. series of moon landers called "Surveyor," Luna-9 did not come down to a completely smooth landing through the use of carefully controlled retro-rockets. The instrument package was carried close to the moon's surface by a rocket, and then jettisoned in a protected ball just before impact. Luna-9 operated for two days, and produced only twenty-seven photographs. Yet these results answered many of the important questions about the moon's surface. Luna-9 was followed by a number of other automatic moon probes in 1966, the most notable of which was Luna-13 in December 1966, which tested the density of the lunar ground with an extendable arm. The implication of these Luna flights, of course, was that a significant amount of exploration could be achieved through automatic devices without risking men's lives.

The year 1966 now seems to emerge as the moment when the Soviet leadership made their difficult decision to abandon the race to the moon. After the triumph of Luna-9, a lively debate ensued among scientists on how best to explore that body: with men, without them, or with some combination of both? To date there is hardly any documentation on this debate, but that it occurred is an established fact. In a dispatch entitled "Automatic Spacemen on the Moon" TASS scientific reporter Larissa Markelova briefly reported that the debate occurred. The article was released to correspondents in Washington in connection with the pioneering flight of Luna-16 in September 1970.[18]

How long the debate went on is hard to say. But there is some evidence that by mid-1966 it had come to an end and that the decision on automatic exploration of the moon was a firm one. The reason for claiming this is an interview that cosmonaut Komarov gave while visiting Japan. On July 11, 1966, the *Asahi Evening News* published the interview. It again asserted the traditional Soviet line: Russia would certainly beat America with men to the moon. But Komarov added, significantly, that Soviet scientists, before dispatching such an expedition, would first send an automatic device to the lunar surface to pick up rock samples. Although he did not go into details, Komarov seemed to be speaking of a device like Luna-16, which did, in 1970, land on the moon and scrape up some lunar soil for the Russians.

The Soviet space program suffered two unexpected blows in 1966 and 1967. Korolyov died on January 14, 1966, and Komarov was killed on April 24, 1967, while testing the Chief Designer's latest spacecraft, Soyuz-1. It is probably no exaggeration to say that Korolyov's passing removed a unique element from the management of the Soviet space program. Thousands of other scientists and technical workers had contributed to the program in the nine years from 1957 to 1966, but Korolyov had been working in aerospace technology from the start—since before the Second World War. He had been invested with unusual positions of power, and exercised both administrative and scientific leadership. After his death, Soviet sources in Moscow acknowledged that he could not be replaced and that his duties would be shared by a number of scientists.

Komarov's death—like the flash fire that consumed an Apollo spaceship and three astronauts in a ground test, January 1967—caused a serious delay in the Soviet manned program. The Council of Ministers and the Communist Party Central Committee immediately appointed a commission, on the very day of the Soyuz-1 catastrophe, to examine in detail what had occurred. Komarov's flight had proceeded normally until he was called down after one day in orbit. Re-entry was routine until the craft began the final stage of descent with the deployment of a main parachute. The parachute did not open properly, and Komarov then deployed an emergency parachute. The shrouds of the emergency parachute became entangled in the main dome and the craft plummeted to earth, killing the cosmonaut on impact.[19]

It took the authorities eighteen months to alter the parachute system and to satisfy themselves of the soundness of the Soyuz ship. If the Soviet Union had been "racing" the United States to the moon, this would have been a very serious and discouraging delay. Finally, at the end of October 1968, the Soyuz flights were resumed. An unmanned Soyuz craft (Soyuz-2) was launched into orbit on October 25, followed the next day by Soyuz-3, which was piloted by a new cosmonaut, Colonel Georgi T. Beregovoi. As the balance of this series unfolded, it became clear that its purpose was to carry out a series of experiments aimed at creating the orbital station that Korolyov had described in general terms in 1963. During the fall of 1967, the fiftieth anniversary year of the Soviet Union, Soviet scientists performed an automatic docking in orbit of two unmanned craft which, at the time, were identified only as Kosmos-186 and Kosmos-188. In January 1969, two manned Soyuz craft (Soyuz-4 and -5) repeated this experiment, and the crews transferred from one cabin to the other by crawling out through airlocks and scrambling through space to the other ship. In October 1969, a record of three Soyuz ships (Soyuz-6, -7, and -8) were placed in orbit, during which various experiments were carried out, including welding in a vacuum inside one of the spacecraft. None of the ships linked, which some Western observers interpreted as an important failure in their flight program. Finally, Soyuz-9, piloted by Colonel Andrian G. Nikolayev and cosmonaut Vitaly I. Sevastianov, set an all-time endurance record of eighteen days in orbit in June 1970.

This version of how the Soviet government came to concede "the moon race" is plausible enough, except for a final mystery. Although Soviet cosmonauts and space scientists stopped making moon landing predictions in the fall of 1968, several notable and inconsistent comments were noticed beginning in the spring of 1969. The first such was made by cosmonaut Vladimir A. Shatalov, a member of the triple-Soyuz flight of 1969. In an interview distributed by the Yugoslav news agency Tanjug, April 9, 1969, he declared that it would take the Soviet Union about six more months to land on the moon. His comments were cryptic, but their thrust seemed to imply that he was speaking of a manned moon landing, although it is not impossible that he had in mind an unmanned flight by an automatic probe such as Luna-16 or -17. Much more explicit was a commentary by cosmonaut Alexei A. Leonov, who declared to a group of Japanese science correspondents in Moscow that the Soviet Union had a manned moon program similar to the U.S. Apollo mission. A dispatch, transmitted by the Japanese Kyodo news agency from the Soviet capital on June 2, caused intense study by Western experts and considerable puzzlement:

MOSCOW, June 2 (Kyodo) —The Soviet Union is planning a manned flight to the moon similar to the Apollo lunar landing being prepared by the United States next month, a Soviet cosmonaut revealed Sunday.

Alexei Leonov, the first man in the world to walk in space in 1965, explained the Soviet lunar project to a group of Japanese science reporters now visiting Moscow.

Leonov said that, if everything goes well, it will be possible for Russia to send a man or men to the moon before the end of this year or early in 1970.

He said he was confident that pieces of rocks picked up on the surface of the moon by Soviet cosmonauts would be put on display in the pavilion during the Japan World Exposition in Osaka in 1970 (Expo 1970).

This is the first indication of any concrete lunar project being pushed by the Soviet Union.

Leonov appeared in military uniform but he responded to questions in a very friendly manner.

At the meeting he indicated that, unlike the Apollo project which sends spaceships directly to the moon from the earth, the Soviet moon project consists of assembling a space station in orbit around the earth.

A lunar probe will be launched from the space station, he said. He said the Soviet Union would get pieces of rocks from the moon either by unmanned or manned vehicles in time for the Expo 1970 exhibition.

This report, which never received official Soviet confirmation, bolstered the view of some Western space experts that the Soviet Union had never really conceded the moon race. To them, the change in tone of Soviet comments in the fall of 1968 was only a tactical maneuver: the Russians realized that they would be outdone by several dramatic U.S. flights, so they decided to stress

the utility of unmanned exploration in which they were
reasonably sure of demonstrating some noteworthy advances. In the meantime, the Western experts held, the
Russians were continuing a determined but secret program to land men on the moon in competition with the
United States.

Bits and pieces of evidence have come forward to
buttress the secret-moon-race theory. Taken together,
these indicators give some weight to the hypothesis,
although none of them is really conclusive. The pieces
of evidence are varied. There were, first of all, before
Khrushchev's ouster, the many comments on beating the
United States to the moon. There was the steady development of a broad program of space exploration that
could support a manned lunar program. There was the
development of a communications network, including
more than a half-dozen specially equipped ships, which
could maintain contact with a manned flight at lunar
altitude. There was the seemingly steady drive to send
a crew on a circumlunar flight, at least until 1967, when
hints of such a spectacular came to an end.

Most convincing of all was the development of a
mysterious booster rocket that Western experts rated as
developing one and a half times as much thrust as the
Saturn rocket, which has taken the Apollo astronauts
to the moon. Soviet scientists acknowledged that they
were seeking to build ever more powerful booster rockets. They disclosed that they had created a rocket, called
the "Proton," which developed three times as much
thrust as the Vostok. But they never confirmed that they
had under development a lunar launch vehicle greater
than Saturn. Academician Leonid I. Sedov, questioned
about this purported moon vehicle during a trip to the

United States in 1968, managed to give an answer that circumvented the question and disclosed nothing.

There is good evidence that this giant rocket does exist, however, and has been detected by U.S. intelligence. Beginning in 1964, James E. Webb, former administrator of the National Aeronautics and Space Administration, revealed the development of the rocket. On a number of occasions, until he retired in 1968, he alluded to it. He described it as possibly developing ten to fifteen million pounds of thrust using conventional, rather than high energy, fuel.

Troubles evidently confounded the builders of the rocket. By mid-1971 it had never flown, even on an experimental mission, although U.S. electronic and satellite intelligence had detected various ground tests. A sensational discovery by this snooping was that, in the summer of 1969, the rocket was devastated on the ground by a fire which broke out in an upper stage.[20] If Leonov's comments to the Japanese seemed to presage the early man-rating of the rocket, Keldysh put a stop to such speculation by the end of 1969. In what may have been an acknowledgment of the troubles that plagued the giant, Keldysh said in October 1969, during a visit to Stockholm: "At the moment, we are concentrating wholly on the creation of large satellite stations. We no longer have any scheduled plans for manned lunar flights."[21]

Possibly this comment by the President of the Academy of Sciences was the final and definitive statement conceding the moon race. Possibly Leonov had spoken at a time when scientists hoped to launch the giant rocket on a successful test flight, and entertained hopes that they could soon man-rate the vehicle. Then, during the summer, the rocket was further grounded by an explosion in an upper stage, and Keldysh made his remark.

I am not inclined to this view. It is true that Leonov's statement is obscure and will require further clarification. But there is a further possible explanation of the booster's mission: the giant booster may have been developed for a variety of purposes that included simultaneously a manned lunar mission, deep planetary probes, and the lifting into orbit of the heavy parts of a manned orbiting platform. Therefore, it was not so very unusual that work continued on the great rocket even after a decision was taken to concentrate on unmanned exploration of the moon.

The theory of a multiple-purpose vehicle is plausible when viewed against the Soviet practice of coordinating, rationalizing, and simplifying wherever possible. Tokaty-Tokaev has remarked that Soviet scientists always tried to construct their major space hardware with a variety of purposes in mind. Thus the Vostok rocket, conceived and developed between 1954 and 1957, has continued in service through 1971 thanks to adjustments and modifications that have made it increasingly powerful. The Voskhod spaceship which carried the first multi-manned crews aloft, was adapted from the original Vostok ship that carried Yuri A. Gagarin.

The Soyuz ship is still another example of this kind of economy. The ship was conceived in three basic parts: a spherical compartment reminiscent of the Vostok spacecraft, a bell-shaped re-entry vehicle with aerodynamic lift qualities, and an instrument section. Western experts believe that the bell-shaped vehicle has been a basic component of the Zond spacecraft that have executed circumlunar flights. The deduction is based on the assumption that Soviet scientists would not design an entirely new re-entry vehicle for circumlunar flights when part of their orbital craft would do just as well. The Zond spacecraft has never been put on public display. But sketches and diagrams of it, which have been published in the Soviet press, corroborate this theory. Photographs of the shipping container of Zond-5, taken as the cargo was transshipped to the Soviet Union from Bombay Harbor, suggest that the vehicle would be of the approximate size of the Soyuz re-entry bell. Statements of scientists that Zond could have carried men on a circumlunar flight also support this theory.

A conclusion emerges from all of this. It seems reasonable to suppose that Korolyov and his technical council conceived the Soyuz craft in the early 1960s before Khrushchev's ouster. (Unmanned flight tests of the ship were detected in 1966 and 1967 under the label of Kosmos-133, -140, -146, and -154.) This period was a time when Khrushchev was intent on keeping the manned moon project alive. The Soyuz ship, then, became a brilliant solution to a variety of problems. It could serve as a basis for constructing a first-generation orbiting platform and it could be used to take men on a flight around the moon, when configured as the Zond spacecraft. It did not have the capability of landing on the moon, however. That task would require further technical developments.

chapter 7

PROPOSALS

When Tsiolkovsky wrote his short story "Beyond the Earth" in the years before the Russian Revolution, he imagined an international team of scientists exploring the moon. His account of the expedition, launched from a pad in the Himalayan Mountains, was remarkably accurate in its major scientific details. The description of the two-man landing party descending on the lunar surface, surveying the landscape, and collecting specimens of moon rock was an amazing preview of the American expeditions which have landed there more than half a century later.

Yet for all his pioneering in physics and jet propulsion, Tsiolkovsky was wrong on a number of points.

The modern moon explorers found no discernible atmosphere, as Tsiolkovsky had anticipated. They uncovered no living world of vegetables and animals whose skins were permeated with chlorophyll. The first expeditions to the moon were not organized by scientific recluses who sought refuge in the Himalayan Mountains. Nor were the first lunar flights cooperative ventures of scientists from different nations, but massive efforts by a superpower which inclined to the belief that it was in a race—a race against another and against time—to reach that satellite.

Why the first flights turned out to be competitive efforts becomes clearer when viewed against the perspective of the last decades. Still, reasonable men are bound to ask what prospects there are for replacing competition with cooperation. Cooperation in exploring the universe is a compelling matter. The size of the task is enormous. The costs are gigantic. The physical risks are unpredictable. Duplication of effort is senseless. And a close integration of efforts would probably contribute something to better understanding on earth.

Cooperation in space, or at least efforts at cooperation, already have a history. The meeting of physicists at James Van Allen's house in Maryland in 1950, at which time the International Geophysical Year was conceived, was a starting point for an unprecedented international effort to explore the earth's surface, its atmosphere, and its relations with the sun. Under the arrangements which evolved, especially in 1954, the United States and the Soviet Union were urged to put their developing rocketry to use in sounding the atmosphere at altitudes that balloons could not reach. In this sense Sputnik was tenuously linked to an international effort. But with the launching of Sputnik-2, which carried aloft the experimental dog Laika, it became evident that the Soviet program had a thrust of its own, independent of IGY purposes. The Soviet Union had secretly developed a serious commitment to explore space. And it set out to do this with tools that had been developed for military and strategic reasons.

The Soviet leaders, no doubt, did not foresee in detail the shock that Sputnik would cause. On the night of October 4, Khrushchev went to bed calmly. Yet in the days to come he would realize a political windfall and take advantage of it. Sputnik was, to be sure, a contribution to the treasury of world knowledge (as TASS said) but it was also a great Soviet triumph. Khrushchev was quick enough to take a bold stance: better to compete in hurling sputniks into space than to up the arms race. This was the Kremlin's posture as the Soviet Union celebrated its fortieth anniversary on November 7, 1957, and through the end of the year.

In due course, the logic of cooperation prevailed and the official Soviet attitude changed in tone. Khrushchev declared that there was enough work to be done by everybody in space, and that Soviet satellites were waiting for the day when American probes would join them. In 1958, the Soviet Union voted "yes" in a United Nations General Assembly resolution calling for the world's potential spacefarers to cooperate, but there the matter rested. At a meeting of the American Rocket Society in Washington in 1959 Soviet scientists explained that space cooperation would have to proceed step by step. They never specified what a first step might be. When Keith Glennan, first administrator of NASA, offered U.S. help in tracking Soviet-manned flights at the end of 1959, the Russians replied they would be in touch if the need arose. They were never in touch.[1]

The years 1960 and 1961 saw little progress in space cooperation. These were years when the Soviet Union, followed by the United States, was hard at work on preparations for launching a cosmonaut into space. The Soviet Union handily beat the United States by successfully launching the first human being in the world into orbit on April 12, 1961. Cosmonaut Yuri A. Gagarin completed one orbit around the earth. Only months later, in August, cosmonaut No. 2, Gherman S. Titov, followed and flew seventeen orbits. The Soviet Union had demonstrated its superiority once again.

Possibly the success of the two manned flights in 1961 over the United States somewhat moderated the Soviet drive in the early space competition. In November 1961, the Soviet Union ended its boycott of the UN Committee on the Peaceful Uses of Outer Space that had already been in existence for two years and began attending meetings. In December, the Soviet delegation to the United Nations voted in the UN General Assembly for Resolution 1721 (XVI), which laid a basis for future space cooperation among the leading space powers. More interesting still was the message of congratulations which Premier Khrushchev sent to President Kennedy on the successful flight of astronaut John H. Glenn, Jr., on February 20, 1962. Riding his Friendship 7 spacecraft through three revolutions, Glenn was the first U.S. astronaut to orbit the earth. Khrushchev wired:

If our countries pooled their efforts—scientific, technical, material—to master the universe, this would be very beneficial for the advance of science and would be joyfully acclaimed by all peoples who would like to see scientific achievements benefit man and not be used for "Cold War" purposes and the arms race.[2]

Khrushchev's thought was appealing. Furthermore, his congratulatory telegram provided a break in the diffident Soviet attitude toward cooperation. Following his inauguration, in his State of the Union message, January 30, 1961, President Kennedy had suggested cooperation with the Soviet Union in a weather-prediction system, a communications-satellite program, in probes to Mars and Venus. He had repeated his interest in cooperation to Khrushchev at their Vienna meeting in June 1961, and returned again to the theme of space cooperation in his address to the UN General Assembly on September 25. Now, following the congratulatory message from the Soviet leader, Kennedy reformulated and renewed his proposals. He replied in a little over two weeks to Khrushchev's telegram—which normally would have required no answer—on March 7, 1962.

The exploration of space is a broad and varied activity, and the possibilities of cooperation are many. I do not intend to limit our mutual consideration of desirable cooperative activities. On the contrary, I will welcome your concrete suggestions along these or other lines.

And he proposed: *

1. The United States and the Soviet Union could establish an early operational weather satellite system.
2. The two countries could track each other's space craft, and train their technicians in each other's space establishments.

*The complete text of his letter is to be found in Appendix B, page 229.

3. The United States and Soviet Union could cooperate in mapping the earth's magnetic field.

4. They could also make joint efforts to develop a satellite communications system which could be made available to the world's nations on a global and non-discriminatory basis.

5. They could pool their efforts and research on space medicine.

Khrushchev likewise answered quickly. On March 20, he dispatched a message in which he described six proposals that responded to and supplemented the Kennedy suggestions: *

1. The United States and Soviet Union should work on the problem of long-distance communication by satellite and should begin by specifying definite opportunities for cooperation in this field.

2. The two countries should cooperate in world-wide weather observation.

3. U.S. and Soviet scientists could participate in a joint program for observing probes launched to Mars, Venus, and other planets of the solar system.

4. It would be desirable to conclude an international agreement to provide aid for search and rescue of downed spacecraft.

5. Soviet scientists could cooperate in the exchange of information on the earth's magnetic field as well as on space medicine.

6. The United States and Soviet Union should try to find a common approach to space law.

The moment seemed propitious. For the first time the two great space nations were talking at the highest

*The complete text of his letter is to be found in Appendix C, page 233.

political level about specific areas where joint efforts might take place. Neither the Kennedy nor the Khrushchev proposals envisaged any intimate integration of the two nation's space efforts. There was no suggestion at this point that the United States and the Soviet Union might pool resources to mount a mixed manned flight, or carry out unmanned planetary probes. Yet the proposals represented a first step which, if successful, could develop mutual trust and might lead to greater coordination.

A second step followed promptly. Dr. Hugh Dryden, the deputy administrator of NASA, was appointed to meet with academician Anatoly A. Blagonravov, who was named Soviet negotiator. Dryden and Blagonravov met in New York, March 27–30, 1962, and again in Geneva, Switzerland, May 29–June 7, 1962, for talks that have been described by the American representatives as congenial, direct, and fruitful. The result of these negotiations was a limited Soviet-American agreement. The Soviet side declined to take up the question of a joint spacecraft-tracking arrangement despite the fact that Khrushchev had indicated willingness in his letter to discuss the matter. The negotiators did agree on a three-part accord under which the two sides would exchange conventional weather data and information obtained from meteorological satellites, share data obtained from satellite observations of the earth's magnetic field, and cooperate on experiments in long-distance communication using the U.S. passive satellite Echo II.

The execution of these agreements over the next few years suffered considerably from technical and bureaucratic inefficiencies.[3] The communications experiments were concluded in February 1964, with several shortcomings. The Soviets declined to transmit signals through Echo II, and limited their participation to receiving signals from the United States. The Russians supplied to the United States useful information about their reception, but the experiments died there. The two nations did not engage in further tests, much less create a global satellite communications system.

For sending weather information the United States and the Soviet Union established a communications channel—dubbed "The Cold Line"—between Moscow and Washington by October 1964. Exchange of conventional weather data began shortly thereafter. However, it was not until September 1966 that data gathered by weather satellites was transmitted, and then only in a hesitant, intermittent manner. Soviet satellite weather data began flowing more regularly after March 1967, but the U.S. side claimed that poor technical quality and late movement made the information practically useless. To improve the situation technical discussions were held in Moscow in July 1968 and in Washington in March 1970. Nevertheless, the two nations were unable to coordinate launchings of operational weather satellites.

Exchanges of magnetic field data, similarly, proved disappointing. Both sides sent a limited amount of information. The Russians forwarded findings of their Kosmos-49 satellite, and the Americans sent data from their U.S. OGO-2 satellite. Shortcomings were reported with regard to regularity, quantity, location, and format of data.

A second cooperation agreement was reached between the United States and Soviet Union in New York on October 8, 1965. This accord involved exchanges on space medicine which had been suggested by both Kennedy and Khrushchev, and, incidentally, also reaffirmed the earlier agreement on weather data. The space medicine accord came into effect rather slowly. A joint editorial board was established and an outline of chapters for a space medicine review was agreed on. However, the Soviet side did not give its assent to the outline for the work until March 1969, and trading of material was delayed several times, beginning in earnest only in January 1970.[4]

At this point it is important to return momentarily to earlier days of Soviet-American space cooperation. In September 1963, President Kennedy made a dramatic offer for a joint U.S.-Soviet lunar mission of which Tsiolkovsky would, undoubtedly, have approved. Kennedy had mentioned the possibility to Khrushchev when he saw the Soviet Premier for the first time at their famous meeting in Vienna. Khrushchev at that time appeared unenthusiastic.[5] His stressing of Soviet superiority implied that the Kremlin felt the United States had a lot of catching up to do. Khrushchev also acknowledged that the secrecy which surrounded the Soviet program could not at that stage be lifted. The Soviet attitude mellowed, however, with the passage of time to the point that bilateral agreements were concluded by Dr. Hugh Dryden and academician Anatoly A. Blagonravov. And there were occasional hints that the Soviet Union might be interested, at least, in exploring the value of joint manned lunar flights as suggested in the famous letter by the British radio astronomer Sir Bernard Lovell to Dryden.

President Kennedy decided to make the offer for a joint lunar expedition in an address to the United Nations on September 20, 1963. He declared before the UN General Assembly that fall:

"Why should man's first flight to the moon be a matter of national competition? Why should the United States and the Soviet Union, in preparing for such expeditions, become involved in immense duplications of research, construction and expenditure?

"Surely we should explore whether the scientists and astronauts of our two countries—indeed, of all the world —cannot work together in the conquest of space, sending someday in this decade to the moon, not the representatives of a single nation, but the representatives of all of our countries."[6]

Kennedy's proposal met a mixed response in the United States, as it did from the Soviet leadership. Some observers suspected that the determination of the Kennedy administration to send an expedition before the end of the decade was weakening. The U.S. Congress amended the NASA appropriations for fiscal year 1964 to restrict the funds for any joint lunar project by requiring prior Congressional approval.

As to Khrushchev, he remained generally noncommittal. In his 1962 letter to President Kennedy, he had noted the obstacles that political considerations occasionally raised for space cooperation. The Soviet Prime Minister referred at least once publicly to Kennedy's joint lunar mission proposal, indicating a vague, general approval. But he never responded seriously or directly. Soviet scientists later acknowledged the potential of such a project, but also echoed concern for the political climate. Lieutenant General Nikolai P. Kamanin, the trainer of the cosmonaut corps, explained the problem succinctly on one occasion. "Before this desirable cooperation can assume concrete form," he said, "more time is needed and above all a favorable political evolution."[7]

The political obstacle to close Soviet-American space cooperation has been cited on many occasions by a variety of Soviet spokesmen. Sometimes the objection is candidly explained, and sometimes it is presented only in a concealed manner. Thus, depending on the circumstances, U.S. involvement in the Vietnam War has been the great barrier; end the war, and cooperation is possible. At other times, Soviet officials insist that the failure of the two superpowers to reach a comprehensive disarmament agreement is the reason why close cooperation in space has not been negotiated. This latter argument has been less used recently as the two nations have engaged in extensive and serious exchanges at the Strategic Arms Limitation Talks in Vienna and Helsinki. Yet the basic problem is clear: The United States and the Soviet Union remain serious political adversaries. The Soviet side would not like to be put in the position of opening up, or subordinating in any major way, an important scientific enterprise to its superpower rival. Conservative attitudes in the Soviet Union on security and secrecy make the problem particularly difficult to conquer. Even the Soviet Union's socialist allies have had to wait a good number of years before close cooperative ties could be achieved.

The Soviet Union has explored the possibilities of space cooperation with its friends and with a few other countries like France and India. Soviet cooperation, on balance, appears to be considerably less extensive than the bonds that the United States has established with its friends. The United States, since 1958, has been cooperating with about seventy nations through nearly 250 cooperative agreements.[8] These arrangements have provided space training for foreign scientists in American universities and in NASA establishments, have involved foreign nationals in tracking U.S. spacecraft in American facilities abroad, have solicited experiments from foreign scientists to be flown aboard American craft, and have provided lunar rock samples for analysis by nearly fifty non-American scientists. To date, the United States has not brought foreign astronauts into its manned flight programs, nor engaged foreign scientists in the planning of future American missions.

Russian spokesmen claim that Soviet space cooperation with foreign countries dates back to the orbiting of Sputnik in 1957. However, this cooperative activity appears to have been merely the tracking of Soviet spacecraft by East European scientists and amateurs, frequently using primitive equipment. A basis for cooperation—not just one-way services—was developed at a Moscow conference on space cooperation only in November 1965. The meeting was attended by Bulgaria, Czechoslovakia, Cuba, East Germany, Hungary, Mongolia, Poland, and Rumania. The deliberations were closed but apparently the parties reached agreements for the supply of component parts and the flying of jointly developed experiments. To implement the arrangements special agencies were created in each of the countries. In the Soviet Union the efforts were coordinated by the Council for International Cooperation in the Exploration of Outer Space. Academician Boris N. Petrov, one of the original members of the 1954 Interdepartmental Commission on Interplanetary Communications, was named the council's director.[9]

Another meeting of these socialist countries was held from April 5 to 13, in 1967. According to a notice published in *Pravda* on April 16, 1967, the nations negotiated a number of agreements relating to space biology, medicine, the launching of joint cooperative satellites, and joint research experiments. Also, the nations considered the possibility of establishing an international satellite communications system.

It was not until December 19, 1968, that the socialist countries launched their first cooperative satellite. This was Kosmos-261, whose mission was reported to be the study of the upper atmosphere, polar lights, and magnetic storms.[10] Beginning October 14, 1969, the Soviet Union and its allies created a new formal series of joint satellites. The series was given the name "Interkosmos." Four Interkosmos satellites have been launched through October 14, 1970. In addition, other Soviet craft, including some of the manned Soyuz flights, have carried various minor components manufactured in Eastern Europe.

Available evidence suggests that the Soviet Union was slow to organize cooperative space ventures, quite apart from the problem of joining the United States. French experience with the Soviet government, further, indicates that the Russian authorities remain reluctant to open up their space program to foreign scientists. Franco-Soviet scientific cooperation is the result of the rapprochement policy pursued by the late President Charles de Gaulle and continued by his successor, Georges Pompidou.

Following de Gaulle's visit to the Soviet Union in June 1966, the two countries established a Franco-Soviet Joint Permanent Commission for developing scientific relations. These relations covered a broad range of activities and originally included a project for launching a French "Roseau" satellite aboard a Soviet booster during the period 1970–1. The Roseau project, however, ran into trouble and was delayed in 1969 until 1972; it was canceled in February 1970. French officials have maintained that the reason for the cancelation was purely financial, but reports have persistently circulated in U.S. space quarters that the French were disappointed that they had not been allowed greater access to the Soviet program. French scientific officials frankly acknowledge that they do not have the same close access to the Soviet program as they enjoy with the United States, and part of the explanation probably remains the continuing Western orientation of the French government. A striking example of the manner in which the Soviet Union continues to hold French scientists at arm's length relates to the laser reflector that was placed on the moon aboard Lunokhod-1. French officials report that when they turned over the reflector to Russian scientists they were told only that it would be placed on the moon sometime before the end of 1970. They were given no advance notice that the automatic probe, which dramatically released a remote-controlled moon rover, would be the vehicle for planting the reflector on the lunar surface.[11]

Nevertheless, Soviet-French relations in space appear to be developing. The two countries have intermittently carried out conjugate soundings by balloons and rockets from the Soviet Arctic village of Sorga and the Indian Ocean archipelago of Kerguelen, nine thousand miles distant along the same line of magnetic force. The French have announced that Soviet rockets will carry into orbit several small technological satellites piggy-back style in the 1970s and that another major satellite project, "Arcade," may replace the canceled Roseau project. On the ceremonial side there have been some notable developments too: de Gaulle and Pompidou have been the only known Westerners ever allowed to visit the Soviet Baikonur Cosmodrome—in 1966 and 1970, respectively.

Another example of Soviet aloofness to Western and American proposals concerns the creation of a world-wide satellite communications system. The project was alluded to by both Kennedy and Khrushchev in their exchange of letters. While the United States and Soviet Union carried out joint experiments with satellite communications, few concrete results were achieved. Meanwhile, the United States forged ahead with preliminary arrangements for the INTELSAT satellite communication system, which was joined by fourteen countries in 1964. INTELSAT in the next few years grew into a global system, handling 90 per cent of the world's long-distance communications through the use of satellites placed over the Atlantic, Pacific, and Indian oceans.

In 1969, the United States convened a conference in Washington to negotiate the definitive arrangements for this system, and to present an opportunity for the system to reduce the preponderant American influence and internationalize itself. In May 1971, the members of INTELSAT successfully completed negotiations for new international arrangements. But the Soviet Union and its East European allies did not join the seventy-seven-nation group. They attended the opening session of the conference in 1969 as observers, and Soviet representative Vladimir Minashin expressed general interest in a single global system if all nations could be assured of equal rights.[12] As the negotiations dragged on into their second year, Soviet interest in the discussion diminished and there seemed no prospect that the Soviet Union would join. Coincidentally, world interest in the Soviet proposal for a global system presented under the name of INTERSPUTNIK in 1968 also faded away.

Nevertheless, the Soviet Union has joined the world's nations in several space agreements. One treaty, modeled on the international agreement on Antarctica, designates space as an area of peaceful scientific research and prohibits the introduction of nuclear weapons on the moon or heavenly bodies. A second treaty establishes procedures for the rescue and return of downed cosmonauts (astronauts) and spacecraft.

Two developments in 1971 were notable. On June 8, the Soviet Union made public a proposed treaty in which the moon would be a non-nuclear area reserved for scientific research. Soviet Foreign Minister Andrei A. Gromyko forwarded the draft of this treaty to UN Secretary General U Thant for discussion. And after seven years of negotiations the Soviet experts agreed to a treaty on liability for damage caused by objects launched into space.

After a decade's efforts at cooperation, one would have to pronounce this arduous history largely a disappointment. The American side, taking advantage of its open space program, has systematically continued the effort to involve the Soviet Union in many different kinds of cooperative arrangements. Pronouncements have been made by President Nixon, Secretary of State William P. Rogers, and on down the line of leadership, to encourage whatever ties may be possible. Recently, the United

States has been urging the Soviet government to join the United States in planning the Viking flights, which are to probe deeply into the solar system in 1973. This effort has been conducted without notable success.[13]

At the conclusion of President Richard Nixon's first year in office, NASA administrator Dr. Thomas O. Paine began a new initiative to engage the Soviet Union. He forwarded to President Keldysh of the Soviet Academy of Sciences the report of Nixon's Space Task Force Group, which examined the problems of space exploration in the coming decade after the Apollo moon flights. On December 12, 1969, Keldysh replied, acknowledging that Soviet-American space cooperation bore only a "limited character." This remark, cryptic as it was, proved to be the beginning of a new phase. While many months passed at an agonizingly slow pace, a concrete response eventually came from the Soviets.

The response was twofold. The Soviet government invited two delegations from NASA, at the end of 1970 and at the beginning of 1971, to Moscow to discuss the possibility of constructing compatible docking systems for Soviet and American spacecraft and to review and improve what little cooperation had been achieved during the previous years.

Dr. George M. Low, acting administrator of NASA, traveled to Moscow for talks of a general nature with President Keldysh in January 1971. During three days of discussions, the two sides held another round of frank and open discussion, culminating in a new agreement. This agreement provided for the exchange of small amounts of lunar samples obtained by the Apollo astronauts and Luna-16. Other parts of the accord called for expanded exchange of information on space biology and medicine, the improvement of exchanges of weather data, consideration by U.S. and Soviet scientists of the results of space research and future objectives, and examination of the world's environment through the use of satellites and conventional means.

The experts' discussion on compatible docking systems began in Moscow in October 1970 and continued with a follow-up conference in Houston in June 1971. Each side grappled with the technical problems in a straightforward manner and at the end of the five-day conference in Texas issued a hopeful communiqué on June 25. Both the Soviet Union and the United States had made progress in determining the technical requirements of compatible systems that they had declared they intended to develop.

The experts also agreed that, after such systems had been installed and tested, a first experiment might be for an Apollo craft to dock with a Salyut module. Subsequently, when the U.S. "skylab" module would be in orbit in 1973 a Soyuz ship could dock with the American orbital station.

The implications of this agreement, which did not attract much notice at the time, were considerable. In the first place, it meant that both nations would develop a capability for carrying out certain kinds of rescue missions in space. Following the catastrophe of Soyuz-11, the appreciation of the risks of space travel was heightened and the value of closer U.S.-Soviet efforts seemed to increase.

Secondly, the ability of Soviet and American spacecraft to dock opened up the possibility in the later 1970s for the development and execution of joint experiments in orbital laboratories. This could now be done without either nation feeling that it had to subordinate a major part of its technology to the desires or plans of the other side.

Once again, hope emerged that the United States and Soviet Union, despite their rivalry on the world stage, might work out means for closer cooperation in space.

chapter 8

THE PRESS

Any correspondent who works in Moscow confronts the Soviet press in his search for information. The press is prolific. It probably supplies the majority of the news that he will dispatch abroad. This reliance on the press has good and bad aspects. Since the Soviet press is a state-controlled institution, the authority of the material actually printed is high. The catch comes in the fact that the press frequently prints only part of the complete story.

It is a Western concept that "news," however unpleasant, should be the most truthful retelling possible of the facts of a particular situation. In Soviet Russia the situation could hardly be more different. There it is openly acknowledged that the press is an instrument that works for a definite master and should contribute to a positive outlook: "News should both serve and help in the resolution of basic problems of Soviet society and Soviet people as they make the gradual transition from socialism to communism. News is agitation by facts. In choosing the subject of news the writer should keep in mind that not all facts nor events must be presented in newspapers."[1]

This is the apt description of Soviet-style "news" by Nikolai G. Palgunov, a former director of TASS. The Western correspondent, or researcher, must simply accept this kind of approach and make as much use as he can of it. There is much that is useful. Moscow publishes an abundance of daily newspapers and journals. Newspapers from the capitals of the fourteen Soviet republics, besides the Russian Soviet Federated Socialist Republic, reach Moscow nearly every day. Hundreds of magazines and specialized publications appear weekly. It would be a half-truth to say that little material is available in Moscow on the Soviet space program or any other subject. There is enough material to drown in. The problem is that information on subjects under censorship is scarce. Since living sources of information are exceedingly difficult to establish, and since the printed word cannot be subjected to interrogation in Western press-conference style, secrets remain well protected.

The machinery for controlling the press is both formal and informal.[2] A primary organ of control is Glavlit, the Chief Administration for Literary and Publishing Affairs. This institution, which existed under the Tsars, was reinstituted by the Bolsheviks in 1922 and has functioned under various names over the years. Glavlit stations representatives at every newspaper who approve each edition. Glavlit authorizes the publication of every journal or book. Its visa numbers appear distinctly on the last pages of every printed work.

Besides this organ, the Section on Agitation and Propaganda of the Central Committee expounds on Party attitudes every several weeks to the top Moscow editors. These editors, by reproducing and elaborating on Party views in their mass circulation newspapers, pass on the cue to lesser editors throughout the country. What the Communist Party daily *Pravda* reports is scanned and studied by millions of workers and managers in their efforts to understand how the Communist Party regards developments. Some of the language is couched in an elliptical manner, and the accomplished reader consequently learns to read between the lines.

Perhaps one of the most efficient tools of press management is self-censorship. Soviet journalists must recognize what the authorities consider news that is "fit to print." There are some general guidelines. News must be upbeat. The Soviet Union is making definite strides toward what its leaders often characterize as "the bright future." Disasters in the Soviet Union should be played down. Political deliberations since Stalin gained control in the late 1920s are traditionally secret. There is no chance that policy considerations in the Central Committee or the Politburo will be published in frank detail, although speeches by leaders to the Central Committee are generally released. With regard to the space program, there is a definite list of secret subjects: the identities of chief designers and managers, the table of organization and responsibility, the exact configuration of the latest rockets and space hardware, projected plans and mission aims, shortcomings and failures which can reasonably be covered up.

The managed press presents the Communist Party with some interesting opportunities and problems with regard to the space program. The Party obviously has the power to encourage certain general attitudes. Yet this power can be eroded over the years by a lack of credibility. Broadcasts received in the Soviet Union from abroad, such as those from the Voice of America and the British Broadcasting Corporation, encourage a credibility gap because they supply details that might normally be restricted. Restriction of information, in the Soviet Union as anywhere else, leads to suspicions of various kinds that are hard to combat.

In October 1960, for example, *Pravda* and *Izvestia* announced the death of Marshal Mitrofan I. Nedelin in an air accident. At the time, Nedelin was one of the most responsible Soviet officials in charge of rocket weaponry. He was both Deputy Defense Minister and commander-in-chief of the Soviet rocket forces. It was not long before rumors, encouraged by the restrictions on news, began to circulate. These rumors were picked up by correspondents and embassy officials in Moscow from Soviet journalists and other sources, and a new picture began to emerge in the West which was much more sensational. According to the unofficial and unconfirmable reports, Nedelin, with other officials, was killed in an explosion of an experimental nuclear-powered rocket. In 1966 (after the sensational story was no longer news), further information on the accident became available in the reports of Oleg V. Penkovsky, the Soviet official and military intelligence officer who became a spy for Great Britain and the United States. He reported:

"Khrushchev had been demanding that his specialists

create a missile engine powered by nuclear energy. The laboratory work concerning such an engine had been completed prior to the forty-third anniversary of the October Revolution in 1960, and the people involved wanted to give Khrushchev a present on the anniversary —a missile powered by nuclear energy. Present during the tests on this engine were Marshal Nedelin, many specialists on nuclear equipment, and representatives of several government committees. When the countdown was completed, the missile failed to leave the launching pad. After fifteen to twenty minutes passed, Nedelin came out of the shelter, followed by others. Suddenly there was an explosion caused by the mixture of the nuclear substance and other components. Over three hundred people were killed."

Today, the biographical notes on Nedelin in Soviet encyclopedias note merely that he "perished in the line of duty" without giving any details or even referring to an air accident.[3]

The launching of Yuri A. Gagarin into space as the world's first cosmonaut was accompanied by some curious press techniques almost completely overshadowed by the astounding achievement and accompanying publicity. The question arose whether the mission was faked in some aspects, or whether the Russian authorities had conducted the flight secretly, publishing "spot" news accounts of developments after it was successfully completed. It is known now that Gagarin's Vostok spaceship was launched from the Baikonur Cosmodrome at 9:07 a.m. Moscow time on April 12, 1961. The cosmonaut landed safely at 10:55 a.m. However, and this can be shown using entirely Soviet documentation, the TASS news agency did not release the first bulletin of the launch until 10:13 a.m.—as the Gagarin flight was

entering into its final phase.[4] At 11:27 a.m., about half
an hour after Gagarin landed, TASS published a dispatch reporting that the ship's retro-rockets had been
activated and that *the descent was beginning.* Finally, at
12:33 p.m., more than an hour and a half after Gagarin
had safely landed, TASS distributed its bulletin reporting his safe return.

One is tempted to accuse the Soviet authorities of
dissimulation. And yet, the real explanation may have
to do with extreme caution in a controlled press—TASS
apparently did not circulate the news of Gagarin's
launch until two conditions obtained: the flight could
no longer be hidden from Western monitors and consequently from Western publicity, and Soviet scientists
were convinced that the flight would end successfully.
This caution may have been the result of a general desire
to maximize success. But it may also have been caused
by unpleasant experience with accidents, which, until
then, had been successfully covered up.

The question of mishaps and catastrophes in the
Soviet space program remains a murky one. Many sensational rumors have circulated about disasters. One
that probably deserves particular attention has been
recalled by newsmen stationed in Moscow at the time
of the Gagarin flight. According to an account published in the Communist *Daily Worker* of London, it was
the son of Sergei V. Ilyushin, not Gagarin, who was to
become the first cosmonaut. The launch was expected
on April 11, 1962, but at the very last moment some
unspecified accident occurred. Ilyushin was said to have
suffered some unspecified injury. For weeks afterward,
reports circulated in Moscow that Ilyushin never quite
returned to normal.

More sensational stories alleged that Soviet cosmonauts actually were lost in space. Observers in various European countries claimed to have monitored conversations between the doomed cosmonauts and ground control. One story asserted that two men and a woman were launched into space and became stranded. As they perished they were reported to have gasped: "Remember us to the Motherland. We are lost; we are lost."[5]

On another occasion, a cosmonaut was supposed to have been launched vertically. The ship reportedly gained extraordinary speed, broke away from the earth's gravitational field, and flew off into space, disappearing into an orbit around the sun.[6]

A decade later there are still no satisfactory explanations for these rumors. They caused considerable embarrassment to the Soviet authorities as witnessed by some of the angry denials that were issued by Alexei I. Adzhubei, chief editor of *Izvestia* and the son-in-law of Nikita Khrushchev, and others. American experts have testified to Congressional committees that the United States possesses no evidence that Soviet cosmonauts have died secretly in flight. When a genuine tragedy did occur, as in the case of cosmonaut Vladimir M. Komarov, the Russians promptly reported the facts of the disaster. Similarly, when Gagarin was killed in an aircraft training flight in 1968 and Pavel I. Belyaev died of internal bleeding a year later, their deaths were quickly announced. The deaths of the three cosmonauts of the ill-fated Soyuz-11 ship were made known within several hours. And yet the spy Oleg V. Penkovsky claimed that several cosmonauts perished in accidents and flights that were covered up.[7]

On January 22, 1969, shots were fired as a triumphal procession of cosmonauts entered the Kremlin through

the Borovitsky Gate. The occasion was the celebration
of the safe return of the crews of Soyuz-4 and Soyuz-5,
who had completed, earlier in the month, the first
manned docking maneuvers by the Soviet Union. Word
of the shooting began to circulate among correspondents.
Finally, a spokesman for the Foreign Ministry reluctantly admitted that such an incident had occurred, and
blamed it on "the work of a schizophrenic." The following day, TASS issued the official Soviet statement:

A provocation took place on Wednesday, when the pilot-
cosmonauts were welcomed in Moscow. Several shots were
fired at the car in which the cosmonauts [Georgi T.]
Beregovoi, [Valentina V.] Nikolayeva-Tereshkova, [An-
drian G.] Nikolayev, and [Alexei A.] Leonov were driven.
   The driver of the car and a motorcycle rider who accompanied the motorcade were wounded. Not one of the cosmonauts was injured. The person who fired the shot was
detained on the spot. The investigation is being conducted.

During the ensuing days, there were a few more
driblets of information. Next day, at the cosmonauts'
press conference on their docking maneuvers, Leonov
commented briefly on the incident but added practically
nothing to the total sum of knowledge. Soviet journalists,
in conversations with Western colleagues, filled in a
few more details. The culprit was apparently a young
man, a Lieutenant Ilyin, on leave from the army, who
had come from Leningrad to Moscow. He had borrowed
the police (militia) uniform of one of his relatives, and
had managed to position himself near the gates by taking advantage of his disguise. Possibly he was trying to
take a pot shot at the top Soviet leaders who were
riding in one of the cars of the motorcade. Then there

followed months of silence during which the case gently faded from public attention into oblivion. On March 20, 1970, the Soviet Supreme Court announced that "a certain Ilyin from Leningrad, born in 1947" had been found insane, not responsible for his actions, and had been committed to a psychiatric hospital for treatment of chronic schizophrenia. The limousine driver died. The motorcycle escort was only slightly wounded.

Two aspects of the official attitude toward this unfortunate event were typical: the reluctance of the authorities to make known disagreeable facts and the failure to provide a convincing explanation for Ilyin's motives. Consequently, it was widely rumored within the Soviet Union and abroad that Ilyin may have been something of a disgruntled Lee Harvey Oswald, who had hoped to assassinate some of the top political figures.

These examples show the negative side of press control. Trying to dampen unpleasant occurrences by restricting information, the authorities actually encouraged rumors, and sensational retellings. The rumors, in turn, tended to discredit the reliability of information being released. On the other hand, these undesirable results—from the point of view of the authorities— must be balanced against advantages that accrue from instructing the press how to report any given situation. The space managers and press authorities hit on one device that has proved very useful in disguising the purpose of a large number of space vehicles. This was the creation of the Kosmos series satellites.

Nikita Khrushchev broke the news of the launching of Kosmos-1 in an excited interjection to a prepared speech in the Kremlin on March 16, 1962. Later the same day, TASS announced that the Soviet Union had begun a new series of soundings whose purpose was to study the upper atmosphere and ionosphere, the earth's cloud systems, charged particles in the ionosphere, and radiation belts. The series was also to study cosmic rays and corpuscular fluxes, and to measure the earth's magnetic field and emissions from celestial bodies. In practice, the series turned out to be much more than that.

Once in orbit, Soviet satellites cannot be hidden from Western monitors or space experts. In fact, all launches are supposed to be registered with the United Nations. In addition, the Goddard Satellite Situation Report in the United States regularly lists all objects placed into orbit. Thus, it is known that between March 16, 1962, and mid-1971 more than four hundred Kosmos satellites were orbited and were subjected to intensive analysis by Western specialists. Among them, the experts report, were several different types of reconnaissance satellites (beginning with Kosmos-4); tests of a fractional orbital bombardment system, sometimes referred to as FOBS (interspersed between Kosmos-139 and Kosmos-289); and tests of a possible inspector-destructor satellite (Kosmos-248, -249, -252, -373, -374, -375). Other members of the Kosmos series have been identified as being military or civilian communications satellites, data-gathering or ferret vehicles, geodesy and navigation satellites.

The Kosmos cover has also been used to carry out life-support experiments (Kosmos-110); to test the Vostok and Voskhod spaceships in precursor flights; to rehearse the automatic docking of the latest-generation Soyuz manned craft without crews (Kosmos-186 and -188).

A number of planetary probes that failed have also

been designated as part of the series. Among these were probes launched at ideal times for exploratory missions of Venus (Kosmos-27, -96, -167, and -359) but which never left earth orbit and decayed; and lunar failures (Kosmos-60, -111, and others). The Kosmos satellites serve as a broad scientific series and a veil around unsuccessful launches.[8]

Another device for cloaking has simply been the silent treatment. A project does not exist until it is announced. Thus, the Russians have never announced that they are working on a mammoth rocket and they refuse to acknowledge that it exists, despite the findings of Western intelligence. When Western monitors turned up evidence during the summer of 1969 that the big rocket had exploded on its testing stand, Soviet authorities remained silent. From their public posture, one would have to conclude that nothing had happened; there was nothing for the Soviet public to get upset about.

One can understand more clearly now the comments of Mstislav V. Keldysh, President of the Academy of Sciences, on the delaying of a Soviet manned expedition to the moon. He said, at a press conference, July 9, 1970, that the Soviet Union was neither planning any immediate flights of man to the moon, nor dropping an important part of its space program. The Soviet Union, he said blandly, had never announced that it was undertaking a manned moon project. Again, since the program was never officially announced, it really did not exist.

The Soviet press, thus, offers space managers and political leaders a convenience. It can be used to smooth away rough patches in performance and dissimulate the military purposes that any large nation developing a powerful space industry would pursue.

The press can also encourage a broad range of attitudes toward space. The Soviet Union is using space for peaceful purposes almost exclusively. The Soviet people can be proud of the homegrown talent that is putting this new technology to use. The world can, and should, be impressed by the taming of the cosmos which the Soviet scientists are achieving through rational and deliberate steps.[9]

There is a definite need, from the Soviet leaders' point of view, to stress the peaceful purposes of space in article after article, statement after statement. The peace offensive and the quest for general and complete disarmament have been an important part of the Kremlin's posture since the 1950s. And occasionally belligerent declarations should not be misinterpreted. There is, true, a close link between the Soviet military and the space effort, but the chairman of the Goskommissiya space agency is a civilian. Soviet authorities would say that the chairman's responsibilities are subordinated to the top political leaders—the military is not dominating the space effort, according to Soviet information, nor trying to militarize space. Khrushchev, it is true, sometimes pointed out that if the Soviet Union could land spacemen at prearranged target areas, it could do the same with enormous warheads. Khrushchev valiantly asserted that this was not a threat, just a technical possibility. And the cosmonauts went on record as saying that Soviet manned spaceships did not carry weapons or bombs on board, and would not do so in the future.

Taking the counteroffensive, the Soviet press has charged that the United States intended to militarize space through the nefarious schemes of the Pentagon generals. Evidence of this was occasionally found in the

American drive to land men on the moon and allegedly
turn it into a military base, or in the U.S. Air Force
project (now scrapped) of creating a manned orbital
laboratory (MOL) for taking observations and carrying
out experiments with military purposes.

The Soviet press regularly has lent its weight to
developing another theme: pride in the motherland's
achievements. Khrushchev talked of the wonderful presents that Soviet scientists gave the nation at the annual
November revolutionary holidays, and the press gave
extraordinary coverage to his words. In the early days,
the press capitalized on the popular enthusiasm for the
pioneering manned launches and captured this jubila-
tion through endless photographs. FIRST IN THE WORLD
was a headline frequently emblazoned across Soviet
newspapers: the first artificial satellite, the first rocket
to the moon, the first man in space, the first group flight,
the first team flight, the first soft landing of an automatic
probe on the moon's surface, the first automatic docking
of spacecraft, the first automatic moon-looper with a safe
return to earth, the first automatic craft to reach the
moon and fly back to earth with lunar samples, the first
lunar rover, the first automatic probe to transmit from
the surface of Venus, the first manned orbital station.
The pride of being first was human and understandable,
besides being generally uplifting and politically desirable.

The coin, also, had another side. Since the early
days of the space age there had been a tendency in the
Soviet press to downgrade or denigrate American efforts.
Khrushchev could mock the first U.S. satellites as hardly
more than "oranges," conveniently disregarding their
scientific discoveries and achievements, which were

occasionally significant. He could boast that the Soviet Union was successful in shooting the moon, while the United States was like a hunter who aimed . . . but missed. His remarks at Sokolniki Park in 1962 on the inadequacies of the American Mercury capsules were much along this line.

Generally speaking, the Russian authorities took the attitude that there was plenty of room for everybody in space and that the United States had only to perform to be recognized. In the meantime, the vast amount of publicity that attended the formulation of plans, the execution of tests, the launching of vehicles in the United States was treated as distasteful and vulgar. With their restrictive views on news, the authorities in the Soviet Union could, for a number of years, pass off American press coverage as indulgent, extravagant, and boastful.

American achievements, however, became more and more a challenge. A necessary adjustment in Soviet public attitudes had to be made at one point, as the United States gained steadily on its objective of a man on the moon "before this decade is out." That point was reached in the fall of 1968. In the Soviet press, an obvious shift occurred. Soviet officials downgraded the idea that men should be risked needlessly in space. When the Americans rounded the moon aboard Apollo 8, congratulations flowed quickly from Moscow to Washington, but the message about automatic devices filtered back loyally too.

Six months later, on July 21, 1969, the United States achieved its goal. Apollo 11 landed on the moon and astronauts Neil G. Armstrong and Edwin "Buzz" Aldrin became the first men to set foot on an extraterrestrial body. The Soviet press gave the event notable publicity.

The moon landing was front-page news throughout the U.S.S.R. Video tapes of the achievement were carried several times on Central Television; laudatory comments were made by Soviet scientists and cosmonauts. President Nikolai V. Podgorny conveyed his congratulations and admiration through astronaut Frank Borman, who was then visiting the Soviet Union on a goodwill tour.

Yet the critical observer might still find in the Soviet press treatment efforts to take the edge off the American feat. Soviet citizens were occasionally reminded of the impressive achievements of Soviet industry. Commentators asserted and reasserted that unmanned exploration of outer space had great achievements behind it and more potential ahead. The press authorities decided against televising the moon landing live as was done in a number of Communist countries, notably Poland, Czechoslovakia, Rumania, Hungary, Bulgaria, and Yugoslavia. Such decisions may have slightly dampened the effect of Apollo 11 in the Soviet Union, and it is a fair question to ask if they were put forward in a spirit of introducing a balancing element during those heady days, or whether they represented an extension of the general tendency of diminishing American exploits.[10]

Another aspect of press treatment concerns the flight of Luna-15. This automatic probe was launched on a mission to the moon shortly before the United States launched Apollo 11 toward its historic landing. True to form, the Soviet authorities spoke cryptically of Luna-15's mission. Unconfirmed rumors circulated that the craft's purpose was to land on the surface, scoop up a few rocks, and return to earth. Should this purpose be fully carried out, the speculation went, and if the United States mission failed or suffered a tragedy, the Soviet

Union would win an enormous last-minute triumph. In any event, newspapers around the world wrote breathlessly of a race to the moon between the men and the robot.

Events favored the United States on July 21, however. Apollo 11 landed and Luna-15 crashed. On analysis later, Western experts doubted that the robot craft of 1969 carried enough fuel to land, pick up rock samples, and return to earth.

Yet Luna-15 left a question: Since a lunar launch opportunity occurs every month, why did the Soviet authorities insist on dispatching the vehicle immediately before Apollo 11? The Russians must have felt that there was a propaganda advantage to this flight: it would divert attention somewhat from the overshadowing Apollo 11 mission and, if disaster engulfed the American spaceship, would provide the opportunity to rehearse the new maxim about the advantages of unmanned exploration of the universe.

Luna-15's mission was officially and vaguely stated to be "to perfect the automatic station's systems on board and carry out further research of the moon and near lunar space." It is a fair presumption that propaganda was also a major purpose, although this cannot be documented. Perhaps someday the secret archives will show a deliberate decision on the part of the authorities to make a last-ditch, even if only partially successful, effort to counter Apollo 11.

Looking backward, it would seem that similar prestige goals have been assigned a number of other Soviet space flights. Khrushchev used to deny that flights were engineered for political purposes. He used to say that, when it came to space flight, scientists dictated the conditions and the dates. Yet the historical record suggests otherwise. Leaving aside the great pioneering firsts—which captured their deserved share of acclaim —there were a number of flights and gestures that showed that the Russians knew how to exploit space for the purposes of enhancing the national image. Khrushchev, in his Washington meeting with President Eisenhower, proudly presented replicas of the medals that Luna-2 scattered on the moon's surface in 1959, thereby rubbing in the Soviet Union's superiority in a dramatic way. In 1963, the Soviet Union orbited a woman cosmonaut, Valentina V. Tereshkova—a feat it shows no immediate intention of duplicating—and then rushed her to the International Women's Congress that was meeting in the Kremlin to become a Soviet showpiece and glamour girl. During the 23rd Communist Party Congress, in March 1966, the scientists launched the automatic Luna-10 probe to the moon. It broadcast from space back to the assembled delegates the "Internationale"—the international Communist hymn.

When Khrushchev was in power he took advantage of achievements in space to drive home his message in a way more eloquent than that of present leadership. Soviet spacecraft, he boasted, were not just launched from Baikonur. They were launched by the great socialist system, in fact, by the "Kremlin Cosmodrome."

This has been a selective picture of how the Soviet press, in conjunction with space and political leaders, combined to create an image of the space effort. On balance that effort was an image of might and pioneering achievement, but it also was an image with flaws, suspicions, and even mysteries. The foreign correspondent operating in this milieu is physically closer to the scene of action than many an observer, but in large measure he is frustrated and cynical. His operating problems are great, aside from the intellectual difficulty of trying to report adequately on hundreds of diverse subjects in a difficult foreign language.

The Soviet authorities have tried, over the years, to influence foreign correspondents much as they control the Soviet press. At best theirs is an imperfect influence, but it is attempted through various devices. The most obvious is an attempt to encourage "self-censorship." Foreign correspondents learn after working in Moscow for a number of months that there exist a number of touchy subjects: the lives of the political leaders and their dignity as statesmen, the intricacies of the Jews' situation in the Soviet Union, the aspirations of the dissident intellectuals and their causes, the secrets of debates and decisions in the high political councils, and, naturally, secrets of a military, aerospace, or scientific value. To go too deeply into these subjects is to court harassment by the authorities in a number of ways that can make everyday life miserable, to invite unpleasant reprimands from the Press Department of the Foreign Ministry, and eventually expulsion. When an expulsion occurs, colleagues inevitably reflect on their friend's fate and draw their own conclusions.

There are other methods of control. It is true that censorship of written newspaper copy was ended in May 1961. Since that time, correspondents have been free to write exactly as they choose—keeping in mind, of course, what the consequences might be. Yet perceptive articles are often difficult to produce because of the correspondents' inevitable reliance on the twists and turns of the Soviet press. Unofficial contacts are difficult to establish and maintain, particularly with responsible workers in government agencies. All news interviews by correspondents with Soviet personalities are supposed to be arranged through the Press Department of the Foreign Ministry, which responds slowly to requests. The TASS news agency will occasionally break a major story on its foreign circuits before making it available to correspondents on the domestic wires. This obliges editors abroad, for lack of expert guidance, to prepare the first stories for the American and foreign press sticking close to the Soviet version. Traditionally the Soviet authorities have opened doors to visiting foreign correspondents (who are often overwhelmed by Russian hospitality), while maintaining a restrictive policy toward the reporters resident in Moscow. In all, the Moscow correspondent who tries to follow the space program labors under a difficult burden, and it is not surprising that he is sometimes less surefooted than he would like to be, even hesitant—as on the occasion when word circulated, at Valentina Tereshkova's wedding reception, that Sergei P. Korolyov was the secret Chief Designer.

PERSPECTIVES

Despite the efforts of Soviet officials to present their space exploration program as a rationally planned effort, which advances step by step without regard to what the United States is doing, there really was "a space race." The Kremlin never officially announced it, but the Soviet Union did participate in it. Particularly during the years when Nikita Khruschev was in power, Russian scientists exerted themselves in various situations to outdo their superpower rival. The then First Secretary of the Soviet Communist Party found exhilaration in the results. As more and more evidence has become available, it seems clearer than ever that the

Kremlin made a conscious effort to orbit Sputnik before the United States' first satellite.

In the years immediately after 1957, Khrushchev capitalized on the political results of the scientific achievement. He used his own moral authority and political suasion to urge on the effort. He understood the political gains of a continuing string of technological spectaculars. The first woman into space symbolized the equal role that Soviet women are supposed to have with Soviet men. It also demonstrated the sophistication of Russian spacecraft and their automatic controls. A first manned expedition to the moon—if accompanied by reliable assurances for the crew's safe return to earth—would also have had, in Khrushchev's opinion, considerable value to the Kremlin at home and around the world.

In the early years, the space program was always producing the unexpected and the exciting. It symbolized the potentialities of which Communist society is supposed to be capable. It brought national euphoria. If properly presented by the press, the space program could boost national morale during years that still remained immensely difficult for the individual man and the frustrated consumer. Uplifting publicity about the pioneering space program could contribute to the strength of the country, and possibly even increase its productivity.

The costliness of space, and the logic of cooperation, were, of course, also obvious to Khrushchev. Yet in the decades of the 1960s there were serious difficulties for this logic. The United States was, after all, the Soviet Union's prime adversary. The United States represented a strategic and military threat to the Kremlin, even if not an immediate one. Furthermore, on the world scene the United States was not infrequently working at cross purposes to Soviet aims. Under no circumstances could the Soviet Union subordinate a major industrial and technological enterprise to its rival. Thus, no close cooperation in the 1960s was possible—and certainly no joint expedition to the moon.

There were other difficulties to cooperation even though Khrushchev and Kennedy might reasonably talk about it. The Soviet Union during the 1960s had its own internal problems in sorting out priorities in space. Khrushchev argued for sensational technological achievements where they could be accomplished with reasonable risks. Some scientists, and possibly some political figures, were more skeptical of benefits of spacefaring. Not until the second half of the 1960s did the Soviet Union under its new leadership of Brezhnev and Kosygin resolve the major problem of how to explore the moon. By the late 1960s, the Soviet Union was definitely committed to a broad program of space exploration in which automatic devices would investigate heavenly bodies; cosmonauts would confine themselves, for the time being, to work on a manned orbital station.

The Soviet Union found that cooperation with its own allies was not such an easy achievement because of the secrecy required by conventional Soviet approaches. To some extent, bureaucratic inefficiencies may also have had a braking effect on joint enterprises.

By the beginning of the 1970s, however, the situation was changed from a decade earlier. The era of the national space spectaculars was over. At great expense the United States had landed men on the moon and demonstrated its ability to mount a series of manned expeditions there. The Soviet Union, through its own intricate internal processes, had decided in favor of unmanned automatic exploration over a broad range of objectives.

Yet various obstacles for joint endeavors remained and they were related to the unalterable fact of super-power rivalry. The Soviet Union and United States remain serious political rivals. Both might *like* to trust the other, but inevitably a high degree of suspicion remains. Different national attitudes compound the problem. The Soviet Union still clings to its secrecy for a variety of military and political reasons. The United States, with its unclassified civilian space program, remains an ardent, and sometimes disappointed, space partner.

Nevertheless, the decade of the 1970s will probably demonstrate the power of the cooperative logic. The renewed Soviet-American space talks in 1970–1 in Moscow and Houston were brought about by a deliberate decision taken at a high level in the Kremlin. To some extent, the tentacles of secrecy were cut in the interests of potentially beneficial projects. Periodic consultations, on the pattern of the talks between acting administrator Low of NASA and President Keldysh of the Academy of Sciences could encourage complementary investigations into the depths of space. The creation of compatible docking systems for Soviet and American spacecraft could have many ramifications for the rescue of stranded spacemen as well as for some joint Soviet-American missions. Such joint missions would undoubtedly relate to observations of the earth and the universe from a manned orbiting station.

The growth of Soviet-American cooperation in space will be arduous and, in the immediate years ahead, unspectacular. It is unlikely that there will be much financial saving in this new cooperation at first, but a new side to the many facets of Soviet-American understanding will develop. In the long run, this understanding with its consequent reduction of suspicions could deliver a few benefits which are, today, difficult to foresee precisely.

APPENDIXES

appendix a

Interview by the Author with
Prof. Dr. G. A. Tokaty-Tokaev at the City University,
St. John Street, London, E.C.1, July 17, 1968

Question: You have described in your writings meetings with Soviet leaders during which you advised on the possibilities of developing long-range rockets. Can you say for what purpose the Soviet leaders were interested in these weapons?

Answer: Which meetings were you referring to? There were several.

Question: I have knowledge only of the meetings which you referred to in your book *Stalin Means War* and in the account that is included in Eugene Emme's *History of Rocket Technology.* By the way, in these accounts the dates differ. In the former you date the meetings in April 1947, and in the latter in March 1947. Which dates are correct?

Answer: The account appeared also in my books *Comrade X* and *Soviet Imperialism,* and in very many articles. However, the dates in question have become mixed up and the error has tended to perpetuate itself. The correct dates are, of course, 14th to 16th April, 1947. I am grateful to you for bringing the matter to my attention: I shall correct the error in future publications.

Question: To return to the aims of the Soviet leaders at the meetings in April 1947 . . .

Answer: Yes, the purpose of the meetings: A good archer is known not by his arrows but by his aim. You see, the U.S.S.R. claims to be a Communist country; and the long-term aim of Communism is the replacement of the capitalist system by the system of Communism. Now, Marx, Lenin, Trotsky and Stalin taught that this aim can only be achieved by means of a socio-political and economic revolution. But such a revolution requires modern armaments. Moreover, the theoretical [sic] aim of Hitler's war against the U.S.S.R. was the destruction of the achievements of the October Revolution and, consequently, the prevention of such revolutions elsewhere. But the Soviet Union came out of World War II as a leading military power, deter-mined to stand up to any new anti-Communist war, to the whole non-Communist world. Above all, this meant standing up to the United States, which by then possessed the B-29 bomber, the A-Bomb, many German V2s, and the leading rocket designers of Germany. The Soviet leaders knew that, until and unless they did something along these lines, they could not stand up to the U.S.A.

To the Russians there was nothing new in rocketry, in general and in a theoretical sense. But during the war the U.S.S.R. had not produced anything like the V2 rocket; therefore the Kremlin leaders were very much worried. . . . In rocketry proper, too, there had been good progress in the Soviet Union. But then there was war. In the beginning our armed forces were smashed. The western part of the country was smashed. We had to dismantle everything and move back to a safe area. We lost every normal condition of work. The war abruptly distorted our work. The U.S.S.R. was losing everything, and the Germans were gaining; the Germans were gaining a great deal from other Europeans. We could do little until the end of the war. We had to move back and then start again at square one. In the end, however, we learned a great deal from the Germans, other European coun-tries, from British and American experience. We were anxious to learn from everyone—which helped.

By the end of World War II, the Soviet Union and the United States constituted two profoundly different worlds. They had to stand up, face to face, as two opposing worlds. Now, remember that the U.S.A. had the long-range B-29 bomber and the A-Bomb: we could be reached by them. But we had neither a bomber capable of reaching the U.S.A. nor the A-Bomb. From a purely military point of view, the situation was really desperate, and hence the line of thought of the Kremlin leaders. There was only one way out: to solve the problems of long-range bombers, rockets and A- and H-Bombs.

This was a direct confrontation of the old enemies—of Com-munism and capitalism. America was the determined leader of the second, and the Soviet Union the determined leader of the first. The so-called "proletarian internationalism" made the U.S.S.R. responsible for communism at large; not just for the U.S.S.R. itself, but throughout the world. For all theoretical and propaganda purposes, this was a great responsibility before the history of Communism. In reality, however, it was a responsibility before the age-long tradition of Russian expansionism. But whatever it was, it required—or demanded—the urgent creation of appropriate material means. What were these means? Well, we began working on an aircraft similar to the B-29. Then, already in 1944, Soviet scientists knew of the V2 and the Sanger project. And it was natural for scientific advisers to call the attention of the leaders to these projects: governments do not make decisions without consultations with advisers.

Question: It has sometimes been suggested in the West that the Soviet Union was interested in its propaganda position while pushing ahead with its space program. Was this question of propaganda image raised in April 1947, at the discussions which you attended?

Answer: No, I do not remember specifically propaganda statements at these meetings. I thought Stalin and his colleagues meant business. But, then, of course, every stick in the world has two ends; a rocket—Soviet or American—is both an effective monster and a propaganda weapon. I also agree that the Soviets love propaganda. But it would be a dangerous illusion to think that Gagarin, Titov, Bykovsky, Nikolayev, Tereshkova, Popovich, Komarov, Belyaev, Feoktistov and Yegorov were nothing more than propaganda. Facts are stubborn things that do not cease to exist because they are painted this or that color.

Having said this, I should now like to focus your attention on something else. You see, the peculiarity of space technology in the U.S.S.R. is that things are designed to fulfill two or more simultaneous functions. Sputnik-1 was a scientific achievement, a heraldic symbol over the gateway into the unknown, a challenge-warning to the capitalist West, an outstanding propaganda drum, etc. And the designers were aware of all these functions. Similarly, the emergence of the purely strategic ICBM was something like a proclamation of the beginning of space exploration, of man's flight around the moon and beyond, of Sputnik, etc. In other words, there are no rockets, spaceships and space pilots devoid of propaganda value, and there has never been a propaganda launching devoid of scientific-technological importance; therefore, he who talks of rockets and sputniks in terms of only propaganda should have his mind examined.

Question: Exactly when was the decision taken to proceed with an ICBM?

Answer: The initial decision was made at the meetings to which we were referring.

Question: In the prewar years, there existed in the Soviet Union several organizations which were concerned with rocket development. I am thinking of GIRD and GDL. These were unified, if I am not mistaken, in 1933 in a national research institute, the so-called RNII. What has happened to these organizations? What has happened, for example, to GDL and to RNII?

Answer: First of all, let me correct you: RNII is not called "national institute." Its exact name is Reaktivnyi Nauchno-Issledovatelskii Institut, which translates into "Reactive Scientific Research Institute." RNII continues to exist, in a greatly enlarged and sophisticated form; and GDL remains a laboratory in it. It is no longer accurate to say, however, that the RNII-GDL complex is the only research establishment dealing with rocketry: no, far from it!

Question: You are in a good position to compare the Soviet and American space efforts. How would you compare them?

Answer: Both programs do exceptionally well. The Americans openly advertise all their efforts—the Soviets openly keep their efforts in secrecy. The Americans try to solve far too many problems—the Soviets make drastic efforts to concentrate on a limited number of goals. The American space program is scattered all over the country; therefore it gives rise to many duplications—the Soviet program tries to keep to the diametrically opposite side of organization.

Question: Is the Soviet space program under the control of the military?

Answer: There is a certain correlation of military and nonmilitary. But this exists also in the United States. Take the Thor rocket—this was used in your country, both as a military weapon and as a space booster. Both you and I can be absolutely sure that THERE ARE NO ROCKETS AND SPACESHIPS IN THE WORLD OF NO OFFENSIVE/DEFENSIVE VALUE; therefore the military of every country is interested in its space programs. I hope this answers the question at least in general terms.

Question: You have written that on April 15, 1947, the Soviet government established a commission to pursue the development of long-range rockets (Pravitelstvennaya Kommissiya po Raketam Dalnego Deistviya). I believe you were a member of this commission. Was it ever acknowledged in the Soviet press?

Answer: No, of course not. No government in the world would publish decisions of this strategic magnitude. Nor were commissions on the development of atomic weapons, tanks, etc., ever publicized.

Question: Can you elaborate on the relations of the scientists and the military?

Answer: Well, let us take GOSPLAN. It is the central planning organ in the Soviet Union. No rocket, aircraft, tank or anything else can be put into production independently or outside the state plans of the GOSPLAN. It works in close cooperation with the Soviet General Staff, therefore there is no duplication of effort. There is a section of the Soviet General Staff which works out in detail military production plans for GOSPLAN.

Question: How did the decision to push ahead with rocketry in 1947 fit into the development of the 1946–50 Five-Year Plan?

Answer: The general goal of this part of the plan was the solution of the problem of rocket launchers by 1950. Then a specialized committee elaborated more precise goals. For example, the improved version of the V2, which was designated the R-14, was planned to reach a serial production stage in the 1947–8 period; therefore, the formation of professional rocket units could be planned for 1950–1.

Question: Did you know Korolyov?

Answer: Yes, I knew him very well.

Question: What did Korolyov do during the war?

Answer: He was continuously working on rockets.

Question: Did he work on rocket engines?

Answer: No, not on engines—not personally. But as a vehicle designer, and as the head of RNII, he influenced the work of rocket engine designers. He worked in the Ministry of Armaments system, while I worked in the Soviet Air Forces system: civilian and military, you might say.

Question: Who is G. V. Petrovich, who writes occasionally—I am thinking of his articles in the *Vestnik* of the Academy of Sciences? Is this name, possibly, a pseudonym for Valentin P. Glushko, another important rocketeer?

Answer: Petrovich? A pseudonym? No, Petrovich is his real name. He works in the GDL in Leningrad.

Question: And Glushko?

Answer: He is working on rocket engines.

Question: How about L. S. Dushkin? What has he worked on?

Answer: Dushkin worked on jet engines as opposed to rocket engines.

Question: What was the purpose of the establishment of the Interdepartmental Commission on Interplanetary Communications of the Academy of Sciences in the fall of 1954, and how did it fit into the organization of the space program?

Answer: The Interdepartmental Commission was established before 1954—I think in 1951—but was announced later. Its task was, and still is, to deal with the problems of general coordination *and* communication with the outside world. Indeed, someone had to speak for the Soviet Union. The problem arose, in particular, in formulating the Soviet position in connection with the International Geophysical Year. But you could not send Korolyov to meetings of the IGY, for example—you had to send someone like academician Blagonravov.

Question: Blagonravov was President of the Academy of Artillery Sciences, which was formed in 1946. Did the academy play an important role in developing rocketry?

Answer: Lieutenant General of Artillery Anatoly Arkadevich Blagonravov, now seventy-four years of age, is a well-known artillery scientist—in ballistics and artillery armaments. For a very long time, he worked in the "Artillery Academy Named After Dzerzhinsky," which was reorganized, after the war, into "Military Engineering Academy Named After Dzerzhinsky." The essence of this reorganization was that it became mainly an academy of rocketry. The reasons: (1) during the 1941–5 war the famous "Katyusha" rockets earned an outstanding reputation; (2) our own and foreign research and development work made it clear that the emergence of medium- and long-range rockets was inevitable; and (3) already by 1947 the U.S.S.R.

was working on the theory and organization of specialized rocket divisions in the Soviet armed forces.

These trends and developments required two kinds of institution: (1) a higher military educational establishment for the preparation of rocket engineers and commanders—the Dzerzhinsky Academy; (2) a central research establishment capable of dealing with the fundamental problems of rocketry and artillery—this task was given to the Academy of Artillery Sciences, which became in due course an Academy of Rocket and Artillery Sciences, headed by Blagonravov.

Question: Did you know Blagonravov?

Answer: Yes, I did. After all, I worked in the Zhukovsky Academy of Aeronautics and he in the Dzerzhinsky Artillery Academy—the two academies cooperated in many fields. Later on, he worked in the Academy of Rocket and Artillery Sciences —and I worked in the field of long-range rockets. I knew a good deal about him, he was one of my early teachers—although not directly, not physically—and I had a high regard for him. I wish him well.

Question: The Soviets are very careful about what they say on their space program?

Answer: They are, indeed. And why should they be careless?! All their statements on research and development are usually approved before they are made. Any paper that is to be presented abroad is approved in advance. Some subjects are simply not discussed at all. For example, the identity of the Chief Designer was never mentioned in the Soviet press. Korolyov's role as Chief Designer was disclosed first by me, in 1960, and acknowledged by the U.S.S.R. itself only after his death.

Question: Would you comment on the timing of the launch of the Soviet Union's sputnik in 1957? Was it intended, do you think, to coincide with the meeting of the IGY preparatory committee in Washington, which was coordinating rocket and satellite research during the IGY? Had it been intended to launch it for the birthday of Tsiolkovsky on September 17?

Answer: The timing of the sputnik launch was too big a piece of cake to play games with. It was launched when it was ready. The Soviet Union tested an ICBM on August 17, 1957, and was ready to fire a sputnik in August. No. Probably not in August. But in September. There was no hurry; it was necessary to check and recheck to make sure that everything was O.K. What really was important was not the dates you mentioned, but November 7, the fortieth anniversary of the 1917 revolution. You will notice that the Soviets did not launch a new Gagarin or anything particularly sensational in 1967, the fiftieth anniversary of the October Revolution: simply because they were not ready. Of course, they achieved an automated space link-up of two satellites, but your question implies, I suppose, that they should have launched something far more impressive. No, no, the Soviets are very careful, very systematic, and would not play with their space reputation.

Question: Alexander Nesmeyanov, the President of the Academy of Sciences of the U.S.S.R., said, I believe, on June 10, 1957, that the Soviet Union had solved the problem of orbiting a satellite. What did he mean by that?

Answer: Yes, I know about that statement by Nesmeyanov. It was made by him on the advice of the Soviet government, and meant precisely what it said. Let me add that at a symposium held at the College of Aeronautics, Cranfield, England, from July 18 to 20, 1957, Professor Boris N. Petrov of the Academy of Sciences of the U.S.S.R. announced quite categorically that A MAN-MADE SATELLITE WOULD BE LAUNCHED IN 1957. May I say that I, for one, "accepted" these statements at once, for good reasons. For, indeed, by 1957, the U.S.S.R. had launched so many rockets that its ability to put a sputnik into orbit was already beyond doubt. The Soviet scientists knew that if a body were launched high enough, at a certain speed, it would stay up there. They also carried out relevant ground simulation experiments.

So, Nesmeyanov's statement meant that they were ready to proceed to practical deeds. They were checking and rechecking the means and techniques. Everything in the U.S.S.R. is studied first theoretically, then experimentally, then produced: all these stages had been completed. The only remaining question was: When?

Question: What effect, would you say, would the death of Vladimir Komarov, the Soviet cosmonaut who died in the descent of his Soyuz spacecraft, have had on the Soviet space program?

Answer: Basically, none. I mean to say that the fundamental goals remain unaffected. But, as you know, there were no manned launchings after Komarov's tragedy. This means that the U.S.S.R. decided not to take further risks. The careful study of all the available materials, including *Aviatsia i Kosmonavtika,* hints that there was a thorough investigation of the technical reasons for Komarov's death, and that they have now been established. One can assume that the main cause of the Soyuz disaster was aerodynamic—which means that the aerodynamic aspects of the vehicle had to be re-examined. In my judgment, the problem has now been solved and an improved version of the Soyuz vehicle will soon resume the flights.

Question: When did these hints appear?

Answer: In this year's issues of *Aviatsia i Kosmonavtika,* for example.

Question: I am grateful for your giving me so much time, but I wonder if I could conclude with a few last questions. You would say, then, that the Russian prewar experience with rocketry constituted a great advantage?

Answer: I would say that the prewar rocketry experience was of unquestionable advantage. It permitted the Soviets to continue along a well-matured general line of development.

Question: I have the impression that the American space program, when compared with the Russians', has the appearance of being more hastily put together and is cramped by the matter of a deadline for landing a man on the moon. Would you comment on this point?

Answer: As I have already said, the Americans do very well. Have you ever been, for example, to the North American Rockwell Corporation—Space Division—in Downey, California? What a wonderful organization it is, I must say. Let me tell you that I have many space acquaintances in America, and wish them well. But I also suspect that President Kennedy was not happy with the rigid formulation that a man should be landed on the moon by the end of the present decade. There was no need to formulate the goal in this way. Moreover, I am a bit surprised that the current president has not been advised to issue an official reformulation of the goal. He could say, you know, that the original deadline is not rigid, but that the effort is going to continue; that the attempt would be made in the next two or three years.

But then, of course, landing a man on the moon is only one of the aspects of the American space program. Let us never forget about the numerous satellites, for example. And in all these fields you, the Americans, do well. I am sure that your Apollo program will reach its goals. But this *could* be achieved without costly duplications. It seems to me that your space effort needs a drastic rationalization. Then there is the problem of the press: it intrudes itself and brings the issue before the public. In the Soviet Union, the situation is different. The decision is made and you get on with the work. It is not written about constantly in the press, it goes ahead secretly.

Question: I assume that when you left the Soviet Union and came to Britain in 1948 you passed on details of Soviet progress to the appropriate authorities. Why, then, was the West so slow to take up the challenge?

Answer: I did not pass on what were regarded, by the Soviets, as secrets. General discussions, yes, but not details. Let me also correct you: I would not say the West was slow. I would say that it was slower than the Soviet Union, but not slow. There were problems in the United States. There was, for example, competition between private companies. Have you seen how keen and how bitter is the competition between firms and corporations? Then there was competition between the branches of the military services. All these complicated the issues.

Question: You mean interservice rivalry?

Answer: Well, I say competition. It is more polite.

Question: In closing, may I ask you about your name? I see some references to "Tokaev" and some to "Tokaty." Which is correct?

Answer: Both are correct. I was born in the north Caucasus and my Caucasian name by birth is "Tokaty." But the Russian version of that name is "Tokaev." The press usually prefers "Tokaev," and I do not mind. As a scientist and educator, I am more known, however, as "Tokaty," and would like to continue under that name. By the way, in many publications I am referred to as "Tokaty-Tokaev," which is also accurate. You can thus choose the name which appeals to you more, and enjoy the game.

Question: Professor Tokaty-Tokaev, I am very grateful for the time you have given me. Thank you very much.

appendix b

Letter of President John F. Kennedy
to Premier Nikita Khrushchev, March 7, 1962

Dear Mr. Chairman:

On February twenty-second last I wrote you that I was instructing appropriate officers of this Government to prepare concrete proposals for immediate prospects of common action in the exploration of space. I now present such proposals to you.

The exploration of space is a broad and varied activity and the possibilities for cooperation are many. In suggesting the possible first steps which are set out below, I do not intend to limit our mutual consideration of desirable cooperative activities. On the contrary, I will welcome your concrete suggestions along these or other lines.

1. Perhaps we could render no greater service to mankind through our space programs than by the joint establishment of an early operational weather satellite system. Such a system would be designed to provide global weather data for prompt use by any nation. To initiate this service I propose that the United States and the Soviet Union each launch a satellite to photograph cloud cover and provide other agreed meteorological services to all nations. The two satellites would be placed in near-polar orbits in planes approximately perpendicular to each other, thus providing regular coverage of all areas. This immensely valuable data would then be disseminated through normal international meteorological channels and would make a significant contribution to the research and service programs now under study by the World Meteorological Organization in response to Resolution 1721 (XVI) adopted by the United Nations General Assembly on December 20, 1961.

2. It would be of great interest to those responsible for the conduct of our respective space programs if they could obtain operational tracking services from each other's territories. Accordingly, I propose that each of our countries establish and operate a radio tracking station to provide tracking services to the other, utilizing equipment which we would each provide to the other. Thus, the United States would provide the technical equipment for a tracking station to be established in the Soviet Union and to be operated by Soviet technicians. The United States would in turn establish and operate a radio tracking station utilizing Soviet equipment. Each country would train the other's technicians in the operation of its equipment, would utilize the station located on its territory to provide tracking services to the other, and would afford such access as may be necessary to accommodate modifications and maintenance of equipment from time to time.

3. In the field of the earth sciences, the precise character of the earth's magnetic field is central to many scientific problems. I propose therefore that we cooperate in mapping the earth's magnetic field in space by utilizing two satellites, one in a near-earth orbit and the second in a more distant orbit. The United States would launch one of these satellites while the Soviet Union would launch the other. The data would be exchanged throughout the world scientific community, and opportunities for correlation of supporting data obtained on the ground would be arranged.

4. In the field of experimental communications by satellite, the United States has already undertaken arrangements to test and demonstrate the feasibility of intercontinental transmissions. A number of countries are constructing equipment suitable for participation in such testing. I would welcome the Soviet Union's joining in this cooperative effort which will be a step toward meeting the objective, contained in United Nations General Assembly Resolution 1721 (XVI), that communications by means of satellites should be available to the nations of the world as soon as practicable on a global and non-discriminatory basis. I note also that Secretary Rusk has broached the subject of cooperation in this field with Minister Gromyko and that Mr. Gromyko has expressed some interest. Our technical representatives might now discuss specific possibilities in this field.

5. Given our common interest in manned space flights and in

insuring man's ability to survive in space and return safely, I propose that we pool our efforts and exchange our knowledge in the field of space medicine, where future research can be pursued in cooperation with scientists from various countries.

Beyond these specific projects we are prepared now to discuss broader cooperation in the still more challenging projects which must be undertaken in the exploration of outer space. The tasks are so challenging, the costs so great, and the risks to the brave men who engage in space exploration so grave, that we must in all good conscience try every possibility of sharing these tasks and costs and of minimizing these risks. Leaders of the United States space program have developed detailed plans for an orderly sequence of manned and unmanned flights for exploration of space and the planets. Out of discussion of these plans, and of your own, for undertaking the tasks of this decade would undoubtedly emerge possibilities for substantive space investigations. Some possibilities are not yet precisely identifiable but should become clear as the space programs of our two countries proceed. In the case of others it may be possible to start planning together now. For example, we might cooperate in unmanned exploration of the lunar surface, or we might commence now the mutual definition of steps to be taken in sequence for an exhaustive scientific investigation of the planets Mars or Venus, including consideration of the possible utility of manned flight in such programs. When a proper sequence for experiments has been determined, we might share responsibility for the necessary projects. All data would be made freely available.

I believe it is both appropriate and desirable that we take full cognizance of the scientific and other contributions which other states the world over might be able to make in such programs. As agreements are reached between us on any parts of these or similar programs, I propose that we report them to the United Nations Committee on the Peaceful Uses of Outer Space. The Committee offers a variety of additional opportunities for joint cooperative efforts with the framework of its mandate as set forth in General Assembly Resolutions 1472 (XIV) and 1721 (XVI).

I am designating technical representatives who will be prepared to meet and discuss with your representatives our ideas and yours in a spirit of practical cooperation. In order to accomplish this at an early date I suggest that the representatives of our countries, who will be coming to New York to take part in the United Nations Outer Space Committee, meet privately to discuss the proposals set forth in this letter.

Sincerely,

John F. Kennedy

appendix c

Letter of Premier Nikita S. Khrushchev to
President John F. Kennedy, March 20, 1962

Dear Mr. President:

Having carefully familiarized myself with your message of March 7 of this year, I note with satisfaction that my communication to you of February 21 containing the proposal that our two countries unite their efforts for the conquest of space has met with the necessary understanding on the part of the Government of the United States.

In advancing this proposal, we proceeded from the fact that all peoples and all mankind are interested in achieving the objective of exploration and peaceful use of outer space, and that the enormous scale of this task, as well as the enormous difficulties which must be overcome, urgently demand broad unification of scientific, technical and material capabilities and resources of nations. Now, at a time when the space age is just dawning, it is already evident how much men will be called upon to accomplish. If today the genius of man has created spaceships capable of reaching the surface of the moon with great accuracy and of launching the first cosmonauts into orbit around the earth, then tomorrow manned spacecraft will be able to race to Mars and Venus, and the farther they travel the wider and more immense the prospects will become for man's penetration into the depths of the universe.

The greater number of countries making their contribution to this truly complicated endeavor, which involves great expense, the more swiftly will the conquest of space in the interests of all humanity proceed. And this means that equal opportunities should be made available for all countries to participate in international cooperation in this field. It is precisely this kind of international cooperation that the Soviet Union unswervingly advocates, true to its policy of developing and strengthening friendship between peoples. As far back as the beginning of 1958 the Soviet government proposed the conclusion of a broad international agreement on cooperation in the field of the study and peaceful use of outer space and took the initiative in raising the question for examination by the United Nations. In 1961, immediately after the first space flight by man had been achieved in the Soviet Union, we reaffirmed our readiness to cooperate and unite our efforts with those of other countries for similar flights. My message to you of February 21, 1962 was dictated by these same aspirations and directed toward this same purpose.

The Soviet government considers and has always considered the success of our country in the field of space exploration as achievements not only of the Soviet people but of all mankind. The Soviet Union is taking practical steps to the end that the fruits of the labor of Soviet scientists shall become the property of all countries. We widely publish notification of all launchings of satellites, spaceships and space rockets, reporting all data pertaining to the orbit of flight, weight of space device launched, radio frequencies, etc.

Soviet scientists have established fruitful professional contacts with their foreign colleagues, including scientists of your country, in such international organizations as the committee of Outer Space and the International Astronautical Federation.

It seems to me, Mr. President, that the necessity is now generally recognized for further practical steps in the noble cause of developing international cooperation in space research for peaceful purposes. Your message shows that the direction of your thoughts does not differ in essence from what we conceive to be practical measures in the field of such cooperation. What then should be our starting point?

In this connection I should like to name several problems of research and peaceful use of space, for whose solution it would in our opinion be important to unite the efforts of nations. Some of them, which are encompassed by the recent UN General Assembly resolution adopted at the initiative of our two countries, are also mentioned in your message.

1. Scientists consider that the use of artificial earth satellites for the creation of international systems of long-distance communication is entirely realistic at the present stage of space research. Realization of such projects can lead to a significant improvement in the means of communication and television all over the globe. People would be provided with a reliable means of communication and hitherto unknown opportunities for broadening contacts between nations would be opened. So let us begin by specifying the definite opportunities for cooperation in solving this problem. As I understood from your message, the U.S.A. is also prepared to do this.

2. It is difficult to overestimate the advantage that people would derive from the organization of a worldwide weather observation service using artificial earth satellites. Precise and timely weather prediction would be still another important step on the path to man's subjugation of the forces of nature; it would permit him to combat more successfully the calamities of the elements and would give new prospects for advancing the well-being of mankind. Let us cooperate in this field.

3. It seems to us that it would be expedient to agree upon organizing the observation of objects launched in the direction of the moon, Mars, Venus, and other planets of the solar system, by radio-technical and optical means, through a joint program.

As our scientists see it, undoubted advantage would be gained by uniting the efforts of nations for the purpose of hastening scientific progress in the study of physics of interplanetary space and heavenly bodies.

4. At the present stage of man's penetration into space, it would be most desirable to draw up and conclude an international agreement providing for aid in searching for and rescuing spaceships, satellites, and capsules that have accidentally fallen. Such an agreement appears all the more necessary, since it might involve saving the lives of cosmonauts, those courageous explorers of the far reaches of the universe.

5. Your message contains proposals for cooperation between our countries in compiling charts of the earth's magnetic field in outer space by means of satellites, and also for exchanging knowledge in the field of space medicine. I can say that Soviet scientists are prepared to cooperate in this and to exchange data regarding such questions with scientists of other countries.

6. I think, Mr. President, that the time has come for our two

countries, which have advanced further than others in space research, to try to find a common approach to the solution of important legal problems with which life itself has confronted the nations in the space age. In this connection I find it a positive fact that at the UN General Assembly's 16th session the Soviet Union and the United States were able to agree upon a proposal of the first principles of space law which was then unanimously approved by members of the UN: a proposal on the applicability of international law, including the UN Charter, in outer space and on heavenly bodies; on the accessibility of outer space and heavenly bodies for research and use by all nations in accordance with international law; and on the fact that space is not subject to appropriation by nations.

Now, in our opinion, it is necessary to go further.

Expansion of space research being carried out by nations definitely makes it necessary to agree also that in conducting experiments in outer space no one should create obstacles for space study and research for peaceful purposes by other nations. Perhaps it should be stipulated that those experiments in space that might complicate space research by other countries should be the subject of preliminary discussion and agreement on an appropriate international basis.

I have named, Mr. President, only some of the questions whose solution has, in our view, now become urgent and requires cooperation between our two countries. In the future, international cooperation in the conquest of space will undoubtedly extend to ever newer fields of space exploration if we can now lay a firm foundation for it. We hope that scientists of the U.S.S.R. and the U.S.A. will be able to engage in working out and realizing the many projects for the conquest of outer space, hand in hand, and together with scientists of other countries.

Representatives of the U.S.S.R. on the UN Space Committee will be given instructions to meet with representatives of the United States in order to discuss concrete questions of cooperation in research and peaceful uses of outer space that are of interest to our countries.

Thus, Mr. President, do we conceive of—shall we say—heavenly matters. We sincerely desire that the establishment of cooperation in the field of peaceful use of outer space facilitate the improvement of relations between our countries, the easing of international tension, and the creation of a favorable situation for the peaceful settlement of urgent problems here on our own earth.

At the same time it appears obvious to me that the scale of our cooperation in the peaceful conquest of space, as well as the choice of the lines along which such cooperation would seem possible is to a certain extent related to the solution of the disarmament problem. Until an agreement on general and complete disarmament is achieved, both our countries will, nevertheless, be limited in their abilities to cooperate in the field of peaceful use of outer space. It is no secret that rockets for military purposes and spacecraft launched for peaceful purposes are based on common scientific and technical achievements. It is true that there are some distinctions here; space rockets require more powerful engines, since by this means they carry greater payloads and attain a higher altitude, while military rockets in general do not require such powerful engines—engines already in existence can carry warheads of great destructive force and assure their arrival at any point on the globe. However, both you and we know, Mr. President, that the principles for designing and producing military rockets and space rockets are the same.

I am expressing these considerations for the simple reason that it would be better if we saw all sides of the question realistically. We should try to overcome any obstacles which may arise in the path of international cooperation in the peaceful conquest of space. It is possible that we shall succeed in doing this, and that will be useful. Considerably broader prospects for cooperation and uniting our scientific-technological achievements, up to and including joint construction of spacecraft for reaching other planets—the moon, Venus, Mars—will arise when agreement on disarmament has been achieved.

We hope that agreement on general and complete disarmament will be achieved; we are exerting and will continue to exert every effort toward this end. I should like to believe that you also, Mr. President, will spare no effort in acting along these lines.

Yours respectfully,

N. Khrushchev

Moscow, March 20, 1962

NOTES

Collected Works, Vol. II (Moscow, 1954), p. 52; also Kosmodemyansky: *Konstantin Tsiolkovsky—His Life and Work,* p. 16.

7. Kosmodemyansky: *Konstantin Tsiolkovsky—His Life and
Work,* pp. 37–8.

8. A. A. Kosmodemyansky: *K. E. Tsiolkovsky—His Life and
Work* (rev., Moscow, 1960), p. 153. The text of the commissars'
decision read as follows:

RSFSR Council of People's Commissars, Moscow, Kremlin,
10. XI. 1921, No. al6085.

The Council of Peoples Commissars at its meeting of
November 9, 1921, having reviewed the question of award-
ing Tsiolkovsky, K. E., an increased, lifelong pension,
decided: "In view of the special services of the scientist-
inventor, specialist in aviation, Tsiolkovsky, K. E., to grant
a lifelong pension."

9. Ibid., pp. 77–80.

10. Kosmodemyansky: *Konstantin Tsiolkovsky—His Life and
Work,* p. 95.

11. Dmitri Ya. Zilmanovich: *F. A. Tsander—Pioneer of
Soviet Rocket Technology* (Moscow, 1966), pp. 3 *ff.*

12. Ibid.

13. N. A. Rynin: *Interplanetary Communications,* Vol. II
(Leningrad, 1928–32), p. 191.

14. V. N. Sokolsky: "Some New Facts About the Works of
Soviet Scientists—Pioneers of Rocket Technology (Through the
Mid-1930s)," paper presented to the XIXth Congress of the
International Astronautical Federation, 2nd History of Astro-
nautics Symposium (New York, October 16, 1968), p. 6.

15. I. A. Slukhai: *Rockets and Traditions* (Moscow, 1965)
(Israel Program for Scientific Translations, Jerusalem, 1968),
pp. 15–16.

16. Rynin: *Interplanetary Communications,* p. 211.

17. F. A. Tsander: *Problems of Flight by Jet Propulsion*
(Moscow, 1961) (Israel Program for Scientific Translations,
Jerusalem, 1964), p. 30.

18. Tsander: *Problems of Flight by Jet Propulsion,* p. 38.

19. A. N. Kiselev and M. F. Rebrov: *Ships Go into Space*
(Moscow, 1967), pp. 26 *ff.*

20. Zilmanovich: *F. A. Tsander—Pioneer of Soviet Rocket
Technology,* p. 174.

21. V. V. Razumov: "History of the Leningrad Group for the Study of Reactive Propulsion," in Academy of Sciences of the U.S.S.R.: *History of Aviation and Cosmonautics, No. 1* (Moscow, 1964), p. 26.

22. Yu. Stvolinsky and I. Churin: "The Rockets of Vladimir Razumov," in *Leningradskaya Pravda* (January 6, 1966), p. 4.

23. Razumov: "History of the Leningrad Group for the Study of Reactive Propulsion," p. 26.

24. I. E. Kulagin: "Work on Rocket Technology in the Leningrad Gas Dynamics Laboratory," paper presented to the XIXth Congress of the International Astronautical Federation, 2nd History of Astronautics Symposium (New York, October 16, 1968), p. 2.

25. S. A. Shlykova: "K. E. Tsiolkovsky's Correspondence with the Jet Scientific Research Institute," in Academy of Sciences of the U.S.S.R.: *From the History of Rocket Technology* (Moscow, 1964) (Israel Program for Scientific Translations, Jerusalem), p. 127.

26. Wernher Von Braun and Frederick I. Ordway: *History of Rocketry and Space Travel* (New York, 1966), p. 60; and Beryl Williams and Samuel Epstein: *The Rocket Pioneers* (New York, 1958), p. 145.

27. Kosmodemyansky: *Konstantin Tsiolkovsky—His Life and Work,* p. 88.

Chapter 2

1. *The New York Times,* November 30, 1961, p. 1.

2. These figures are drawn primarily from two sources—Academy of Sciences of the U.S.S.R. (Institute of History): *A Short History of the U.S.S.R.* (Moscow, 1965), pp. 245 *ff.*; and Department of State: *Foreign Relations of the United States, Diplomatic papers: The Conference of Malta and Yalta, 1945* (Washington, 1955), p. 705.

3. Isaac Deutscher: *Stalin, A Political Biography* (London, 1949), p. 548.

4. Department of State: *Foreign Relations of the United*

States, Diplomatic papers: The Conference on Berlin (The Potsdam Conference), 1945, Vol. II (Washington, 1960), p. 853.

5. *Pravda,* February 10, 1946, p. 1.

6. Vasily S. Yemelyanov: "Kurchatov as I Knew Him," in *Yunost,* No. 4 (April 1968), p. 88.

7. Vasily S. Yemelyanov: "Kurchatov as I Knew Him" (concluded), in *Yunost,* No. 5 (May 1968), p. 91.

8. V. L. Sokolov: *The Soviet Use of German Science and Technology* (New York, 1955), pp. 10 *ff.*

9. On the role of the Academy of Artillery Sciences in developing rockets for military purposes, see Academy of Sciences of the U.S.S.R.: *The Great Soviet Encyclopedia* (Moscow, 1949), p. 562; U. S. Senate, Committee on Aeronautical and Space Sciences: *Soviet Space Program* (Washington, May 1962), pp. 64–5.

10. See H. Grottrup: "Aus den Arbeiten des Deutschen Raketen-Kollektivs in der Sowiet Union," in *Raketentechnik under Raumfahrtforschung* (April 1958), Heft. 2, p. 58.

11. Ibid.

12. Interview with Grigory A. Tokaty-Tokaev, July 17, 1968.

13. G. A. Tokaty-Tokaev: "Soviet Rocket Technology," in Eugene M. Emme: *History of Rocket Technology* (Detroit, 1964), p. 276.

14. G. A. Tokaty-Tokaev: "Foundations of Soviet Cosmonautics," *Spaceflight* (London), Vol. X, No. 10 (October 1968), p. 344.

15. Tokaty-Tokaev's most informative accounts of the Kremlin meetings I found to be in *Comrade X* (London, 1956), pp. 311 *ff.* and "Foundations of Soviet Cosmonautics," pp. 343 *ff.* His article in *Spaceflight* is, in fact, a slightly expanded version of a taped interview prepared for the National Air and Space Museum of the Smithsonian Institution in Washington, D.C.

16. Tokaty-Tokaev: *Comrade X,* p. 327.

17. Ibid., p. 329.

18. Interview with Grigory A. Tokaty-Tokaev, July 17, 1968.

19. Walter Sullivan: *Assault on the Unknown* (New York, 1961), pp. 20–8.

20. International Council of Scientific Unions: *Annals of the International Geophysical Year* (London, 1958), p. 456.

21. *Izvestia,* September 24, 1954, and *Vechernaya Moskva,* April 16, 1955. *Vechernaya Moskva,* Moscow's evening newspaper, described the commission on April 16, 1955:

A permanent Interdepartmental Commission on Interplanetary Communications has been established under the Astronomy Council of the U.S.S.R. Academy of Sciences. The Commission must coordinate and direct all work concerned with solving the problem of mastering cosmic space.

Comrade A. G. Karpenko, scientific secretary of the Commission, reported the following to a *Vechernaya Moskva* correspondent:

"The problem of realizing interplanetary communications is undoubtedly one of the most important tasks among those which mankind will have to solve on the way to conquering nature. The successful solution of this task will become possible only as a result of the active participation of many scientific and technological collectives. It is precisely for the unification and guidance of those collective efforts of research workers that the permanent Interdepartmental Commission on Interplanetary Communications has been established at the U.S.S.R. Academy of Sciences. The Commission is headed by academician L. I. Sedov, and is composed of outstanding scientists-physicists, mechanical engineers, physicists, and others—among them academicians P. L. Kapitsa and V. N. Ambartsumyan, corresponding member of the U.S.S.R. Academy of Sciences P. P. Parengo, Doctor of Physics and Mathematics B. V. Kukarkin, and others.

"One of the immediate tasks of the Commission is to organize work concerned with building an automatic laboratory for scientific research in space. Since it is outside the limits of the atmosphere, such a cosmic laboratory, which will revolve around the earth as its satellite for a long time, will permit observations of phenomena that are not accessible for investigation under ordinary terrestrial conditions. Thus, biologists will be able to study conditions of life when the force of gravity is absent; astro-physicists will be able to observe the ultraviolet and x-ray spectra of the radiation of the sun and the stars and to learn much about the processes which are taking place on these bodies. Such laboratories will enable radio physicists to study more completely the processes which take place in the ionosphere and to determine the most advantageous conditions for establishing radio communications with the spaceships of the future. Geophysicists and geographers will be able to provide more accurate results in forecasting the weather, and further help northern navigation by photographing at any moment the dispositions of ice floes in the Arctic Ocean and of the clouds in the atmosphere.

"The creation of a cosmic laboratory will be the first step in solving of the problem of interplanetary communications and will enable our scientists to probe more deeply into the secrets of the universe."

22. Dr. Clifford C. Furnas: "Birthpangs of the First Satellite," in *Research Trends* (Cornell Aeronautical Laboratory, Inc.), spring 1970, pp. 15 *ff*.

23. *Pravda*, January 23, 1957, p. 1.

24. Aleksei Ivanov: "How It Began," *Izvestia*, October 4, 1967, p. 4.

25. *The New York Times*, October 5, 1957.

26. *The New York Times*, October 9, 1957.

Chapter 3

1. G. A. Tokaty-Tokaev's report, "Soviet Space Technology," has been published in *Spaceflight* (London), Vol. V, No. 2 (March 1963).

2. Aerospace Information Division, Library of Congress: *Top Personalities in the Soviet Space Program* (Washington, D.C., May 26, 1964).

3. Nikita S. Khrushchev: *For Victory in Peaceful Competition with Capitalism* (New York, 1960), p. 197.

4. Vladimir Orlov in *Pravda*, August 26, 1961.

5. *Izvestia*, June 20, 1961, p. 1.

6. Ibid.

7. Ibid.

8. Alexander P. Romanov: *Designer of Cosmic Ships* (Moscow, 1969), pp. 101–2 and footnote 1, p. 102.

9. Documentation on the State Commission for the Organization and Execution of Space Flight can be found in a variety of

sources—Aerospace Information Division, Library of Congress: *Management of the Soviet Space Program* (Washington, D.C., October 24, 1963); A. N. Kiselev and M. F. Rebrov: *Ships Leave for Space* (Moscow, 1967), p. 318; and Alexander P. Romanov: *Cosmodrome, Cosmonauts, Cosmos* (Moscow, 1966).

10. Romanov: *Designer of Cosmic Ships*, p. 64.

11. Ibid., p. 86.

12. Romanov: *Cosmodrome, Cosmonauts, Cosmos*, pp. 113–14.

13. For a description of Tikhonravov's background and possible role in the Soviet space program, see Aerospace Information Division, Library of Congress: *Top Personalities in the Soviet Space Program*, pp. 17 *ff*.

14. Romanov: *Designer of Cosmic Ships*, pp. 154–5.

15. B. V. Shipov: *National Rocket Building* (Moscow, 1967), p. 82.

16. Based on conversations with officials at the National Aeronautics and Space Administration, Washington, D.C., fall 1970.

17. *Izvestia,* June 13, 1962.

18. Theodore Shabad in *The New York Times,* June 26, 1966.

Chapter 4

1. Published documentation on Chief Designer Sergei P. Korolyov is becoming more available. A number of useful books and articles are—Alexander P. Romanov: *Designer of Cosmic Ships* (Moscow, 1969); P. T. Astashenkov: *Academician S. P. Korolyov* (Moscow, 1969); P. T. Astashenkov: "The Rising Current," *Aviation and Cosmonautics* (Moscow), issues of October through December 1968.

2. Romanov: *Designer of Cosmic Ships*, pp. 24 *ff*.

3. *Pravda,* September 30, 1967, abridged interview of 1963 with TASS correspondent Alexander P. Romanov.

4. Romanov: *Designer of Cosmic Ships*, p. 35.

5. Ibid., p. 41.

6. Ibid., pp. 45 *ff*.

7. Evidence that Korolyov was confined after the war has been presented by Leonid Vladimirov, a Soviet journalist who defected in the 1960s to Great Britain. In his article, "From

Sputnik to Apollo," published in the Russian émigré journal *Posev,* September 1969, Vladimirov specifically refers to Korolyov's detention. Astashenkov, the Soviet biographer of Korolyov, makes an allusion to this period in the academician's life in his volume *Academician S. P. Korolyov,* pp. 98 *ff.*

8. Astashenkov: *Academician S. P. Korolyov,* p. 99.

9. Romanov: *Designer of Cosmic Ships,* p. 51.

10. Astashenkov: *Academician S. P. Korolyov,* p. 112.

11. G. A. Tokaty-Tokaev: "Foundations of Soviet Cosmonautics," *Spaceflight* (London), Vol. X, No. 10 (October 1968), p. 340.

12. Astashenkov: *Academician S. P. Korolyov,* pp. 102 *ff.*

13. S. P. Korolyov: "Practical Significance of the Scientific and Technical Proposals of K. E. Tsiolkovsky in the Field of Rocket Engineering," in Soviet National Association of Historians of Natural Science and Technology: *History of Aviation and Cosmonautics,* 1965, Vol. IV (translation of the National Aeronautics and Space Administration, November 1967).

14. Romanov: *Designer of Cosmic Ships,* p. 53.

15. Astashenkov: *Academician S. P. Korolyov,* p. 119.

16. Romanov: *Designer of Cosmic Ships,* p. 64.

17. Astashenkov: *Academician S. P. Korolyov,* pp. 141 *ff.*

18. Ibid., pp. 162 *ff.*

19. Yuri A. Gagarin: *Road to the Stars* (Moscow, 1962), p. 122 (English edition).

20. Romanov: *Designer of Cosmic Ships* (Moscow, 1969), pp. 109 *ff.*

21. Yuri A. Gagarin: *Road to the Cosmos* (Moscow, 1969), p. 321.

22. Astashenkov: *Academician S. P. Korolyov,* pp. 200 *ff.*

23. *Pravda,* January 16, 1966.

Chapter 5

1. Nikita S. Khrushchev: *Khrushchev in America* (New York, 1960), p. 105.

2. Nikita S. Khrushchev on May 10, 1962, quoted in Richard W. Porter and Charles S. Sheldon: *A Comparison of the United States and Soviet Space Programs* (Washington, D.C., 1965), p. 32.

3. Interview with Nikita S. Khrushchev by Serge Groussard, correspondent of *Le Figaro*, March 19, 1958, in Nikita S. Khrushchev: *For Victory in Peaceful Competition with Capitalism* (New York, 1960), p. 197.

4. Nikita S. Khrushchev during an appearance at the National Press Club, Washington, D.C., September 16, 1959, in Nikita S. Khrushchev: *Khrushchev in America*, p. 29.

5. Interview with Nikita S. Khrushchev by the Hearst Task Force, November 24, 1957, in the *Journal-American*, November 25, 1957, pp. 1 *ff*.

6. Nikita S. Khrushchev in a speech at Minsk, Byelorussia, January 22, 1958, in Nikita S. Khrushchev: *For Victory in Peaceful Competition with Capitalism*, pp. 34–5.

7. Nikita S. Khrushchev in a speech in East Germany, July 9, 1958, in Nikita S. Khrushchev: *For Victory in Peaceful Competition with Capitalism*, pp. 530–1.

8. Interview with Nikita S. Khrushchev by the editor of *Dansk Kolkestyre* on January 4, 1958, in Nikita S. Khrushchev: *For Victory in Peaceful Competition with Capitalism*, pp. 23 *ff*.

9. Nikita S. Khrushchev: *The Construction of Communism in the U.S.S.R. and the Growth of Agricultural Development* (Moscow, 1963), Vol. V, pp. 95–6.

10. Interview with Nikita S. Khrushchev, *The New York Times*, September 8, 1961, pp. 1–10.

11. Nikita S. Khrushchev in *Pravda*, January 15, 1960, p. 1.

12. Vannevar Bush: *Modern Arms and Free Men* (New York, 1949), p. 86.

13. Eugene M. Emme: *History of Rocket Technology* (Detroit, 1964), p. 75.

14. John F. Kennedy, "Memorandum for Vice-President," April 20, 1961.

15. Khrushchev: *The Construction of Communism in the U.S.S.R. and the Growth of Agricultural Development*, Vol. IV, p. 397.

16. Documents of the 22nd Communist Party Congress, pp. 269 *ff*.

17. *The New York Times*, January 7, 1963.

Chapter 6

1. Academician Leonid I. Sedov stated in Turin, Italy, according to United Press International on December 28, 1968, that the Soviet Union was stressing automated probes and would not send men to the moon in the near future. The general attitude of the Soviet Union toward "the space race" and the Apollo 8 achievement was stated at the end of 1968 by President Mstislav V. Keldysh of the Academy of Sciences and Moscow Radio in two rather interesting commentaries,

According to a press conference transcript published in *Izvestia*, November 6, 1968, p. 4, Keldysh said:

> If two scientists are working in similar fields, is there a competition between them? Undoubtedly there is to a certain extent. But it is important that this should not be a determining factor, because if one were to devote all to the benefit of the competition, to the benefit of a race, then one might forget about science.

Moscow Radio said in a broadcast beamed to the United States, December 28, 1968:

> To be quite frank, we would naturally have liked Soviet astronauts on board a Soviet spacecraft to have been the first to fly so far out into space. But we realize that space exploration is a job of unprecedented scope, not to be approached with the yardstick of athletic competition. We certainly feel no jealousy toward Americans or begrudge them this lead in making this distant space flight. We have faith in the space program worked out by Soviet scientists and engineers. This program was a success in the past and is well under way now. We are confident it will be no less successful in the future.

2. For an assessment of the Soviet space program, several studies by Charles S. Sheldon, chief of the Science Policy Research Division of the Library of Congress, are helpful. These are: *United States and Soviet Progress in Space: Some New Contrasts* (Washington, D.C., January 12, 1971); *United States and Soviet Progress in Space: How Do the Nations Compare?* (Washington, D.C., January 31, 1970); *United States and Soviet*

Rivalry in Space: Who Is Ahead, and How Do the Contenders Compare? (Washington, D.C., March 31, 1969); and *Review of the Soviet Space Program,* report of the Committee on Science and Astronautics, U.S. House of Representatives (Washington, D.C., 1967).

3. Charles S. Sheldon: *United States and Soviet Progress in Space: Some New Contrasts* (Washington, D.C., 1971), p. 35.

4. Ibid., p. 11.

5. *The New York Times,* December 8, 1965, pp. 1 *ff.,* an interview with James Reston.

6. *Izvestia,* July 11, 1970, p. 2.

7. U.S. Library of Congress, Aerospace Information Division: *Future Lunar Missions* (Washington, D.C., January 8, 1964).

8. Gherman S. Titov, quoted in *Ultimas Noticias* (Mexico City), October 23, 1968, and published in *The New York Times,* October 24, 1968, p. 5.

9. U.S. Library of Congress, Aerospace Information Division: *Future Lunar Missions,* pp. 95 *ff.*

10. Ibid., p. 98. Keldysh commented in an interview broadcast by Radio Prague, October 16, 1963.

11. Letter of Sir Bernard Lovell to the author, September 21, 1970.

12. Leonid Vladimirov: "From Sputnik to Apollo," in *Posev* (Munich), September 1969, pp. 47–51.

13. *Aviation and Cosmonautics* (Moscow), 1968, p. 376.

14. Moscow Radio and Television, October 12, 1964.

15. Ibid., October 13, 1964. Also reproduced in Yuri A. Gagarin: *Road to the Cosmos* (Moscow, 1969), pp. 329 *ff.*

16. Vladimirov: "From Sputnik to Apollo," pp. 47–51.

17. *Pravda,* October 22, 1968, pp. 1 *ff.*

18. Larissa Markelova: "Automatic Spacemen on the Moon"; circulated in Washington to news correspondents following the flight of Luna-16 by the Novosti Press Agency of Moscow.

19. See the description of the accident given by cosmonaut Konstantin P. Feoktistov at a news conference October 23, 1968, at the Apollo News Center, Houston, Texas.

20. *Aviation Week and Space Technology* (New York), November 17, 1969, p. 26.

21. *The New York Times,* October 26, 1969, p. 1.

Chapter 7

1. Hearings before the Committee on Aeronautical and Space Sciences, U.S. Senate, March 11, 1970, Part 3, p. 935.

2. Committee on Aeronautical and Space Sciences, U.S. Senate: *Soviet Space Programs,* 1962–5 (Washington, D.C., December 30, 1966).

3. Hearings before the Committee on Aeronautical and Space Sciences, p. 934.

4. Ibid.

5. Theodore C. Sorensen: *Kennedy* (New York, 1965), p. 529.

6. Committee on Aeronautical and Space Sciences, U.S. Senate: *Soviet Space Programs,* p. 441.

7. Ibid., p. 442.

8. Hearings before the Committee on Aeronautical and Space Sciences, pp. 920, 939 *ff.*

9. TASS News Agency, Moscow, dispatch of October 14, 1970.

10. Soviet News, Press Department of the Soviet Embassy in London, issues of October 21, 1969, p. 30, and July 7, 1970, p. 9.

11. Conversation with scientific officials at the French Embassy in Washington, November 24, 1970.

12. *The New York Times,* March 1, 1969.

13. Hearings before the Committee on Aeronautical and Space Sciences, p. 935.

Chapter 8

1. Nikolai G. Palgunov: *The Foundations of Information in the Newspaper: TASS and Its Role* (Moscow, 1955), p. 35.

2. For a penetrating account of the operations of the Soviet press, see Mark W. Hopkins: *Mass Media in the Soviet Union* (New York, 1970).

3. Oleg Penkovsky: *The Penkovsky Papers* (New York, 1965), pp. 337–9. See also *Encyclopedic Dictionary* (Moscow), Vol. II, p. 86, for a standard biographical note on Marshal Nedelin.

4. Alexander P. Romanov: *Cosmodrome, Cosmonauts, Cosmos* (Moscow, 1966), pp. 10 *ff*. For the official launch and landing times of Vostok-1, see A. A. Blagonravov, *et al.*: *The Achievements of the U.S.S.R. in the Study of Cosmic Space* (Moscow, 1968), p. 433.

5. Committee on Science and Astronautics, U.S. House of Representatives: *Review of the Soviet Space Program* (Washington, D.C., 1967), pp. 51 *ff*.

6. Ibid.

7. Penkovsky: *The Penkovsky Papers*, pp. 339–40.

8. Committee on Science and Astronautics, U.S. House of Representatives: *Review of the Soviet Space Program*, pp. 63–70; also Michael Stoiko: *Soviet Rocketry* (New York, Chicago, San Francisco, 1970), pp. 110 *ff*.

9. Committee on Aeronautical and Space Sciences, U.S. Senate: *Soviet Space Programs, 1962–5* (Washington, D.C., December 30, 1966); see Chapter II: "Political Goals and Purposes of the U.S.S.R. in Space."

10. Hearings before the Committee on Aeronautical and Space Sciences, U.S. Senate, March 11, 1970, Part 3, p. 1016: "Worldwide Treatment of Current Issues—Special: Apollo 11."

SELECTED BIBLIOGRAPHY

INDEX

Academy of Sciences of the U.S.S.R., Institute of Natural History and Technology: *Pioneers of Rocket Technology* (selected readings from the works of Nikolai I. Kibalchich, Konstantin E. Tsiolkovsky, Friderikh A. Tsander, and others). Moscow, 1964.

Aerospace Information Division, U.S. Library of Congress: *Future Lunar Missions*. Washington, D.C., 1964.

Aerospace Information Division, U.S. Library of Congress: *Management of the Soviet Space Programs*. Washington, D.C., 1963.

Aerospace Information Division, U.S. Library of Congress: *Top Personalities in the Soviet Space Program*. Washington, D.C., 1964.

Astashenkov, P. T.: *Academician S. P. Korolyov*. Moscow, 1969.

Blagonravov, A. A., *et al.*: *Soviet Rocketry: Some Contributions to Its History*. Israel Program for Scientific Translations; Jerusalem, 1966.

Emme, Eugene M.: *History of Rocket Technology*. Detroit, 1964.

Frutkin, Arnold W.: *International Cooperation in Space*. Englewood Cliffs, N.J., 1965.

Gagarin, Yuri A.: *Road to the Cosmos: Notes of a Flyer-Cosmonaut of the U.S.S.R.* Moscow, 1969.

Gatland, Kenneth W.: *Astronautics in the Sixties*. London, 1962.

Hearings before the Committee on Aeronautical and Space Sciences, U.S. Senate, International Space Cooperation, Part 3. Washington, D.C., March 11, 1970.

Khrushchev, Nikita S.: *For Victory in Peaceful Competition with Capitalism*. New York, 1960.

Khrushchev, Nikita S.: *The Construction of Communism in the U.S.S.R. and the Growth of Agricultural Development*. Moscow, 1963.

Kiselev, A. N., and M. F. Rebrov: *Ships Go into Space*. Moscow, 1967.

Kosmodemyansky, A. A.: *Konstantin Tsiolkovsky—His Life and Work*. Moscow, 1956. Rev. edn., Moscow, 1960, as *K. E. Tsiolkovsky—His Life and Work*.

Krieger, F. J.: *Behind the Sputniks: A Survey of Soviet Science*. Washington, D.C., 1958.

Porter, Richard W., and Charles S. Sheldon: *A Comparison of the United States and Soviet Space Programs*. Washington, D.C., 1965.

Romanov, Alexander P.: *Cosmodrome, Cosmonauts, Cosmos*. Moscow, 1966.

Romanov, Alexander P.: *Designer of Cosmic Ships*. Moscow, 1969.

Slukhai, I. A.: *Rockets and Traditions*. Moscow, 1965.

Sheldon, Charles S.: *Review of the Soviet Space Program*. Washington, D.C., 1967.

Sheldon, Charles S.: *United States and Soviet Progress in Space: How Do the Nations Compare?* Washington, D.C., 1970.

Sheldon, Charles S.: *United States and Soviet Progress in Space: Some New Contrasts*. Washington, D.C., 1971.

Sheldon, Charles S.: *United States and Soviet Rivalry in Space: Who Is Ahead, and How Do the Contenders Compare?* Washington, D.C., 1969.

Sokolsky, V. N.: *Russian Solid-fuel Rockets*. Moscow, 1963.

Stoiko, Michael: *Soviet Rocketry: Past, Present and Future*. New York, 1970.

Tokaty-Tokaev, G. A.: *Comrade X*. London, 1956.

Tokaty-Tokaev, G. A.: "Foundations of Soviet Cosmonautics," *Spaceflight* (London), Vol. X, No. 10 (October 1968).

Tokaty-Tokaev, G. A.: *Stalin Means War*. London, 1951.

Tokaty-Tokaev, G. A.: "Soviet Rocket Technology," in Eugene M. Emme: *History of Rocket Technology*. Detroit, 1964.

Zilmanovich, Dmitri Ya.: *Pioneer of Soviet Rocket Technology —F. A. Tsander*. Moscow, 1966.

A Note About the Author

Nicholas Daniloff was born in Paris, France, in 1934. He studied in France, England, and the United States and joined United Press International in 1959. He worked in the Moscow Bureau of UPI from 1961 to 1966, and since 1967 has been a State Department correspondent for the agency in Washington, D.C. Mr. Daniloff lives in Washington with his wife and two children.

A Note on the Type

The text of this book was set on the Linotype in a type face called Life. Brought out by the German type foundry of Ludwig and Mayer in 1964, Life is an adaptation by Francisco Simoncini of Times Roman, the popular British type face. Designed especially for use in a newspaper, Times Roman is widely appreciated for its legibility. Life, with its angular points exaggerated to compensate for loss of detail in reproduction, seems ideally suited for both text and newspaper settings.

This book was composed by Cherry Hill Composition, Pennsauken, New Jersey, and printed and bound by Haddon Craftsmen, Scranton, Pennsylvania. Typography and binding design by Anthea Lingeman.